Cellular Communications Explained

From Basics to 3G

Cellular Communications Explained
From Basics to 3G

Ian Poole

AMSTERDAM • BOSTON • HEIDELBERG • LONDON • NEW YORK • OXFORD
PARIS • SAN DIEGO • SAN FRANCISCO • SINGAPORE • SYDNEY • TOKYO
Newnes is an imprint of Elsevier

Newnes is an imprint of Elsevier
Linacre House, Jordan Hill, Oxford OX2 8DP
30 Corporate Drive, Burlington, MA 01803

First edition 2006

British Library Cataloguing in Publication Data
A catalogue record for this book is available from the British Library

Library of Congress Cataloging-in-Publication Data
A catalog record for this book is available from the Library of Congress

ISBN-13: 978-0-7506-6435-5
ISBN-10: 0-7506-6435-5

For information on all Newnes publications visit our web site at www.newnespress.com

Printed and bound in UK
06 07 08 09 10 10 9 8 7 6 5 4 3 2 1

Contents

Preface

From relatively small beginnings in the 1970s and 1980s, the cellular telecommunications industry has grown to be one of the most important areas of electronics today. It is now one of the major drivers for technology, and its influence has been a significant factor in moving semiconductor technology to provide solutions that are lower cost, use less power, and provide increased levels of processing power and flexibility. Additionally, it has driven forward the level of integration of RF circuitry, combining many of the RF circuits onto a single chip with the digital processing circuitry. All of this seemed a far-off dream in the early 1980s, when the first commercial cellular networks were being launched and rolled out.

With the development progressing swiftly from the first analogue systems to the second-generation digital and then the third-generation high-speed multimedia-compatible systems, there has been a very swift increase in the capability of the systems. Even while the 3G systems were being rolled out, supplementary enhancements were added to improve the performance still further, creating what were termed 3.5G systems. In addition to this, new technologies were being developed and trails were being run to investigate the technologies for the fourth-generation systems. Now it is possible to surf the Internet, send emails, store video clips, send pictures and even talk to others.

The speed of the developments, the diversity of technologies required and the number of companies involved make this a fascinating and lively arena. Nevertheless the technologies are challenging, and this resulted in many delays to the roll-out of the 3G systems. However, the need to communicate, and the increasing number of capabilities offered in today's phones, mean that the industry is set to continue its growth.

Despite the global nature of cellular communications and international roaming, there is a considerable number of systems in use. While GSM is undoubtedly the major standard, passing the 1 billion subscriber barrier in February 2004 and still continuing to grow, there are several other second-generation systems in use. With the pace of the 3G roll-out increasing, this has brought further standards.

This book has been written to provide a basic understanding of the major cellular technologies in use around the globe. The aim is to provide a grounding in the basic concepts and principles, and then to move on to describe the individual standards, including some information about the analogue systems but focusing on the more widely used technologies, including GSM with GRPS and EDGE, cdmaOne (IS-95), CDMA2000, and UMTS (wideband CDMA) – the successor to GSM.

As with any book, it has been necessary to elicit the help of others in obtaining the information and also checking the text. Thanks are due to Marios Agathangelou, Brian Gardner, Mike Henley and Phil Medd. They all contributed useful suggestions and comments regarding the manuscript, which were invaluable in the preparation of the work. Many thanks for your hours of work and helpful comments.

Ian Poole
July 2005

Introduction to cellular telecommunications

Mobile phone technology is now a major aspect of today's life, both business and personal. Instant access to people, wherever they are, is now an accepted part of today's culture. Business requires that people, whether at home or abroad, remain in constant touch, and this is now possible through the development of the mobile phone. In their private lives people have also come to depend on mobile phones, initially using them sparingly and only in cases of emergency, but they have now become an accepted part of everyday life. Many people do not have a traditional landline, and rely only on the mobile for their telecommunications requirements.

Beginnings

Before work started on developing the mobile phone itself, there were many technologies that needed to be in place. Obviously the work of the early pioneers, including Volta, Ampère, Galvani and many more who established the foundations of electricity, was paramount. However, electricity was not used as a means of communication for a number of years. Long-distance communication was generally by written message carried by a courier. Other systems were also used, but these were either mechanical or crude in nature – for example, a network of bonfires was set up along the South of England to warn of the invasion by the Spanish Armada, and in 1792 Claude Chappe devised and installed some semaphore towers in France, for which the word *telegraph* was coined. However, it took the discovery of electromagnetism by Hans Christian Oersted before viable electrical systems could be developed. One of the first schemes to be tested was developed by Wheatsone and Cooke. This used a variety of needles to point to the relevant letter. Although a trial system was installed between Paddington Station in London and Slough to the west of London, its use was never widespread because it required five wires – and insulated wire was very expensive at the time. Nevertheless, the imagination of the public was fired when a murderer was arrested as a result of this telegraph. A man named John Tawell had escaped from the scene of the crime in Slough, travelling on the train to London. A description of Tawell was sent ahead to Paddington Station by telegraph, and he was arrested on his arrival there.

It took an inventive American named Samuel Morse to devise a viable system. An unlikely inventor, Morse was an artist – one of the finest that America has ever produced. On a return ship journey from Europe he heard about the discovery of the electromagnet, and started to think of ways it could be used in an electrical communication system. On his return to the USA, his painting and teaching activities took precedence and the idea lay dormant. However, he enlisted the help of some others to speed the development, and the system for opening and closing a circuit to send a series of coded characters started to come together. Realizing that they would need backing from large organizations if they were to be able to install the system, they took the idea around several organizations but there was little interest and the group split. Morse persevered, and eventually managed to secure a grant from the US Congress to install a trial system from Washington to Baltimore. On 24 May 1844, he sent the famous message 'What hath God wrought'. This started one of the largest communications revolutions ever, and the Morse system (see Figure 1.1), with its accompanying Morse code, entered the history books. The idea quickly spread, not only through the USA, but also worldwide. In Britain, for example, the telegraph enabled the government in London to communicate with people in the colonies around the world.

The next major event was the development of the telephone. After the invention of the telegraph, a number of people worked on transmitting sound over wires. In 1857 an Italian-American named Antonio Meucci developed a primitive telephone system but, coming from a poor background, he was unable to obtain any financial backing. The traditionally acknowledged inventor of the telephone was a Scot named Alexander Graham Bell.

Bell conceived his idea in the summer of 1874, which was to generate a 'speech shaped electric current'. To achieve this, in June 1875 Bell tried a system whereby a stretched parchment membrane, with one end of a ferro-metallic reed attached to the centre, was placed over the pole of an electromagnet. Sounds caused the reed to vibrate over the electromagnet and generate a 'speech shaped electric current'. However, the results were a little disappointing, as the sounds it produced were very muffled. The following year Bell tried a new system. This consisted of a damped reed receiver and a new type of transmitter or microphone – an idea that had previously been tried by Elisha Gray in his telephony work. The device consisted of a diaphragm, attached to which was a metal wire which hung into a dilute acid solution; the sounds from the diaphragm would move the wire up and down in the acid, thereby changing the resistance of the circuit. The first telephone message took place on 10 March 1876 when Bell spoke to his assistant, saying 'Mr Watson, come here, I want you'. Bell had spilled some acid over his clothes and wanted some assistance. With this success the telephone system was born, and it soon started to make a large impact.

Although originally Bell was credited with the invention of the telephone, in recent years the American Congress has given that honour to Antonio Meucci. Meucci had filed a law suit against Bell, but did not have the means to support it and died before it came to court.

With the telephone system established, the next major development was that of wireless (or radio) technology. James Clerk Maxwell was the first to deduce mathematically the existence of electromagnetic waves. It then fell to Hertz to prove their existence, relating them to Maxwell's

Figure 1.1 A camelback Morse key with sounder dating from around 1860. These keys were used by Morse telegraphers in the USA. The shape of the key which is used to give it balance for ease of operation also gave rise to its name.

equations, although a number of other people before him had undoubtedly seen effects of radio waves.

Initially Hertzian waves (as they were first known) were seen as little more than a scientific novelty. However, a young Italian named Marconi did much to exploit them and apply them to practical uses for communication. Seeing their potential for enabling communication between ships, he first approached the Italian navy; when he was turned down, he came to Britain with his mother (who was of Irish stock) and started to develop his ideas here. He successfully demonstrated communications over increasing distances, finally, in 1901, transmitting a signal across the Atlantic.

Marconi concentrated on the marine market, as did many others. Here, wireless was the only means of communication over long distances, and it was especially valuable in sending distress messages. A station was set up in the South Goodwin Lightship, not far from Dover in the UK, and a link between the lightship and the South Foreland lighthouse enabled a number of emergencies to be reported – including one where a ship named the *S.S. R.F. Matthews* collided with the lightship.

Radio technology continued to develop, especially with the introduction of the thermionic valve (Figure 1.2). This enabled signals to be amplified and processed more effectively. Until this point receivers had been severely limited by a lack of sensitivity. Also, transmitters were often spark transmitters that spread their energy over a wide range of frequencies. The introduction of the valve enabled oscillators using a single frequency to be built.

The two world wars gave impetus to radio technology development, but the next major step forward took place after the Second World War. A research programme had been organized in the USA, by Bell Laboratories, to investigate the possible use of semiconductors in electronics. Teams were set up to work on different areas of semiconductor-related research. One team, headed up by Shockley and including Bardeen and Brattain, started to investigate a three-terminal field-effect device. Initially unable to make it operate, they switched their efforts to other areas. Eventually they managed to develop a device consisting of two back-to-back diodes, which was the first transistor – a point-contact device that provided gain. After having the idea, they tried it and it worked first time. A week later, on the day before Christmas Eve 1947, they demonstrated it to executives at Bell.

While the transistor was being developed, others at Bell Laboratories were looking ahead to other ideas. In 1947, D. H. Ring put forward a proposal for a radio system that would use a number of lower power transmitters in 'cells' to enable the re-use of frequencies – a critical element if a large number of people were to be allowed access to a system. The proposal even mentioned the need for a method of 'handing over' the mobile station from one cell to the next as it moved along. However, Ring's document does not state how this might be achieved. Moreover, radio and electronics technology had not advanced sufficiently for the idea to be implemented and, as a result, it lay dormant for several years.

Meanwhile, transistor technology started to advance. The original point-contact transistor was not reliable and, only a few weeks after the invention of the first transistor, Shockley proposed the

Figure 1.2 An example of an early thermionic valve. This 'R' valve dates from shortly after the First World War.

junction transistor. With further developments in semiconductor technology, improved methods of processing the materials and of manufacturing were developed. As a result, transistors became cheaper to produce, their performance improved and they became more reliable, leading to an increase in their use. The field-effect transistor that Bardeen, Brattain and Shockley had tried to develop also came to fruition, and was to play an important part in one of the next major developments – that of the integrated circuit.

There had been a number of projects set up to investigate how electronic circuits could be made smaller and more reliable. However, the development of the integrated circuit has been attributed to two individuals. The first was Jack Kilby, then a young engineer working for Texas Instruments. Having insufficient leave, he had to work during the company shutdown. As there was little call on his time from others, and all the equipment he needed was available, he started work on developing a small oscillator on a single chip of silicon. Working on his own, he made the first circuits work successfully on 12 September 1958. The second, Robert Noyce, working for Fairchild, reasoned that it was nonsensical to make a large number of individual transistors on a wafer, cut them up to make separate transistors and then reassemble them when equipment was constructed. Noyce applied this concept, and set down many of the foundations on which today's integrated circuit industry is founded.

With many of the enabling technologies in place, the scene was set for mobile phone technology to start to become a reality. There had been a number of intermediate steps along the way.

Mobile radio was already in use. The first walkie-talkies had been made in the USA by Motorola in 1940, and were still very heavy (35 lb, or about 16 kg), but they enabled the military to have radio communications on the move. After the war, mobile car telephones were introduced – the first from AT&T in St Louis, Missouri, USA, in 1946. The service was very successful, and soon spread to twenty-four other cities. However, these telephones were effectively two-way radios linked to the ordinary phone network. The services used a transmitter–receiver station located in the centre of the relevant city and, accordingly, had limited range. Also, owing to the limited number of frequencies available, there was a waiting list many times longer than the number of people who were connected. Services were also set up in other countries around the world, with the same problems of waiting lists longer than the number of users.

Seeing the popularity of these services, and realizing their potential, the idea of a cellular system like that previously suggested by Ring resurfaced. AT&T lobbied the FCC (Federal Communications System) in the USA repeatedly between 1958 and 1968, and finally the FCC agreed to set aside some frequencies for an experimental system. As a result, a radio telephone system employing frequency re-use was set up aboard a train in 1969. A total of six channels in several zones were used along the route, which spanned over 200 miles, with the system under computer control.

Meanwhile, in the late 1960s and early 1970s a number of countries started to consider seriously the possibility of a cellular telecommunications system. In Japan, for example, the Nippon Telegraph and Telephone Company proposed a nationwide cellular system at 800 MHz. Ideas also started to move forward in Finland. Then, in December 1970, in the USA, the Bell Telephone Laboratories submitted a patent proposal.

It took until 1975 before the FCC gave approval for Bell to start a trial system, and two more years before it was allowed to operate. Not surprisingly, the development of a new technology cost a very significant amount; many millions of dollars were spent, and eventually systems started to be seen. In fact, the first commercial development cellular telephone system began operation in May 1978 in Bahrain. Although relatively simple in some respects, the system had two cells and about 250 subscribers. However, development in the USA moved ahead very swiftly, and two months later, in July 1978, the Advanced Mobile Phone Service (AMPS) commenced operation around Chicago. Initially the system was trialled using Bell employees, but in December of that year paying customers started to use the system. It took until 1983 before full commercialization of the system took place in the USA; however, the first mobile phone system to be launched commercially was the Nordic Mobile Telephone (NMT), which was launched in 1979 using a band of frequencies at 450 MHz. This was the result of cooperation between engineers and different companies across Scandinavia. Figure 1.3 shows a 1992–1996 Nokia 101 handset for NMT 900 networks.

Development in many parts of Europe followed on behind the USA. A system known as Total Access Communications System (TACS), developed by Motorola, was used in many countries. In the UK, licences were awarded in 1985. Two companies were given licences; one company was partly owned by the previously state-owned British Telecommunications (BT), and the other was called Racal Vodaphone. Owned by Racal Electronics plc, this company was later floated as a separate company to become Vodafone, now one of the world's largest mobile phone companies.

Figure 1.3 Nokia 101 handset for NMT 900 networks, 1992–1996 (reproduced courtesy of Nokia).

Naturally, cellular telecommunications technology spread around the world to many countries and several other standards were introduced. Although analogue systems worked well they had some drawbacks, and ideas for digital systems were forming. One of the first was a European initiative which started its life as the Groupe Speciale Mobile. Its name was later changed to the Global System for Mobile Communications, although the initials GSM were retained. Initial work for this started in 1982. A total of twenty-six telecommunications companies within Europe cooperated on the development of the new system, and it commenced operation with frequencies in the 900-MHz band in mid-1991. The same basic system is also used at 1800 MHz, where it was initially known as the DCS 1800 (Digital Communications System) or GSM 1800, and in North America at 1900 MHz, where it was called the PCS 1900 (Personal Communications System) or GSM 1900. Spectrum was also later released for GSM at around 850 MHz in North America.

With GSM established, one of the features that had originally been included, more for engineering use, was the possibility of sending text messages. The idea was introduced, and after the ability to send messages from phones on one network to those on another had been enabled, this service took off in a large way. Called the Short Message Service (SMS), it was initially thought to be an interesting development; however, its use rose rapidly, especially among young people who found it a cheap way of communicating using their phones. In 2004, over 45 billion messages were sent each month. February 2004 also saw the one-billionth GSM subscriber connected.

In North America, an equivalent system to GSM was introduced. Again, its aim was to be able to make more efficient use of the available spectrum. In outlying areas, especially, there was not the same pressure on the spectrum as in many of the cities; accordingly most users were very satisfied with their analogue systems and the take-up was less than in Europe. Nevertheless, the North America TDMA system, which operated on a 30-kHz channel spacing, was rolled out in many areas. A very similar system was also used in Japan, where it was known as Pacific or Personal Digital Cellular (PDC).

Meanwhile, in the USA, a company named Qualcomm proposed a system based on a spread spectrum technique, previously used mainly for military covert activities. The first issue of the specification, IS-95, was available in draft format in 1993, with September 1995 seeing the first deployment of the system, by Hutchison Telecom in Hong Kong, followed shortly afterwards by SK Telecom in South Korea. The system was given the trade name cdmaOne™, and soon started to spread in both the USA and Asia Pacific regions.

As the 1990s came to a close, the cellular phone industry was booming. Industry analysts reasoned that people would want to use far more data services as they saw a significant rise in the use of the Internet. Existing systems were not able to support sufficiently fast data services, and new systems were sought. The first step on the way was known as the General Packet Radio System (GPRS), and its enhanced system as Enhanced Data rates for Global Evolution (EDGE). These systems were dubbed 2.5G, as they were a development of the second-generation system. Similar requirements were also placed on cdmaOne. Although it was able to provide low data rates, its specification was upgraded to support more data activity.

However, the main goal was a fully third-generation system. Three 3G systems emerged. In Europe, a system known as the Universal Mobile Telecommunications System (UMTS), using wideband CDMA (W-CDMA), appeared. The complexity, coupled with the very high prices paid for the licences and the downturn in the global industry, led to delays, although systems did start to be launched. The first system in the UK was launched by '3' on 3 March 2003 (3–3–03) – an apt launch date for a 3G service. Other followed some time after, and the pace of roll-out soon increased. Additionally, many more phones became available (Figure 1.4).

In the USA, a system known as CDMA2000 was adopted. This provided an evolutionary path from cdmaOne through to the full 3G standard with backward compatibility. The first commercial launch of CDMA2000 was in October 2000 in South Korea, with the CDMA2000 1X system. Other countries and operators soon followed as the possibilities of what was termed a 3G service were realized, although the 1X system did not have the full data rates of future services. The full 3G data rates arrived with CDMA2000 1xEV-DO, a data-only service that was first launched in January 2002 when SK Telecom (Korea) initiated a network and provided the opportunity for users to download clips of the Olympic Games being held there that year. The next stage of the CDMA2000 development evolution is the CDMA2000 1xEV-DV, which is a data and voice standard supporting full 3G data rates.

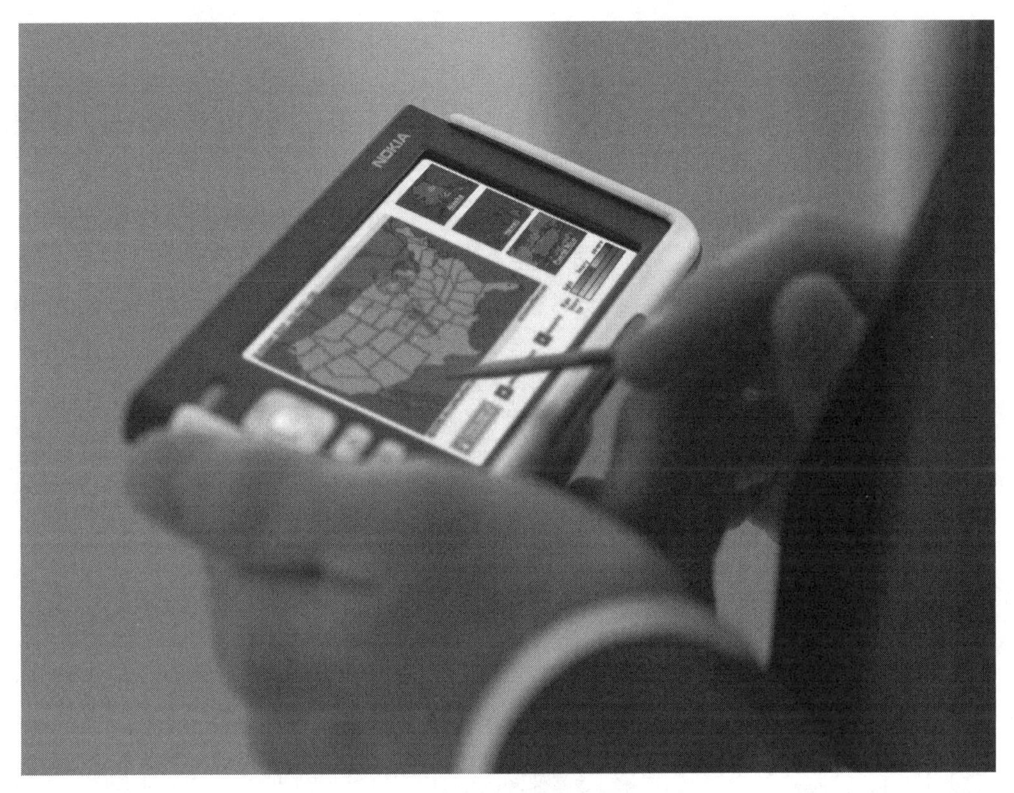

Figure 1.4 The Nokia 770 Internet tablet (reproduced courtesy of Nokia).

A third system, known as Time Division Synchronous CDMA (TD-SCDMA), was developed in China. Although the Chinese rolled out both UMTS and CDMA2000, they were keen also to have a system developed in China. Using a TDD approach, where the transmission and reception were undertaken on the same channel but split in time, this approach has a number of advantages to offer under some circumstances. However, its development is somewhat behind that of UMTS and CDMA2000, and it will not be addressed in the later chapters.

Overview of the systems

Since the introduction of the first cellular telecommunications networks, many different standards or systems have been used, all with their own abbreviations. They also have different specifications, such as channel spacing and access technology (i.e. the means for supporting a large number of users), and they offer different levels of capability. Before progressing to the way in which cellular phones work, and the details of the different major systems, it is worth taking time to overview the different major systems that are (or have been) in use around the globe.

As already mentioned, there are the three generations of system, generally known as 1G, 2G and 3G to denote the different generations. The first-generation systems were characterized by the fact

Table 1.1 Summary of the major cellular telecommunications systems and their capabilities.

System	Generation	Channel spacing	Access	Comments
AMPS	1G	30 kHz	FDMA	Advanced Mobile Phone System; this analogue system was first developed and used in the USA.
NAMPS	1G	10 kHz	FDMA	Narrow band version of AMPS, chiefly used in the USA and Israel, based on a 10-kHz channel spacing.
TACS	1G	25 kHz	FDMA	An analogue system used originally in the UK and based around 900 MHz, this system spread worldwide. After the system was first introduced, further channels were allocated to reduce congestion, in a standard known as Extended TACS (or ETACS).
NMT	1G	25 kHz	FDMA	Nordic Mobile Telephone. This analogue system was the first system to be widely used commercially. Initially based on 450 MHz and later at 900 MHz, it was used chiefly in Scandinavia but was adopted by up to thirty other countries.
NTT	1G	25 kHz	FDMA	An analogue system used in Japan, operating at frequencies in the region of 900 MHz.
C450	1G	20 kHz	FDMA	System used in Germany, operating in the region of 450 MHz.
GSM	2G	200 kHz	TDMA	Originally called Groupe Speciale Mobile, the initials later stood for Global System for Mobile communications. It was developed in Europe and first introduced in 1991. The service is normally based around 900 MHz, although some 850-MHz allocations exist in the USA.
DCS 1800	2G	200 kHz	TDMA	1800-MHz derivation of GSM, this is also known as GSM 1800.
PCS 1900	2G	200 kHz	TDMA	1900-MHz derivation of GSM, this is also known as GSM 1900.
US-TDMA	2G	30 kHz	TDMA	This system, sometimes called North America Digital Cellular or US Digital Cellular, was introduced in 1991 and is also known by its standard number (IS-54), which was later updated to standard IS-136. As it is based on a TDMA system, it is normally referred to as US-TDMA or just TDMA in the countries where it is used. It is a 2G digital system that was designed to operate alongside the AMPS system.
PDC	2G	30 kHz	TDMA	Pacific Digital Cellular. A system similar to NA-TDMA, but used only in Japan.
GPRS	2.5G	200 kHz	TDMA	General Packet Radio Service. A data service that can be layered onto GSM. It uses packet switching instead of circuit switching to provide the required performance, and data rates of up to 115 kbps are attainable.

EDGE	2.5/3G	200 kHz	TDMA	Enhanced Data rates for GSM Evolution. The system uses a different form of modulation (8PSK) and packet switching, which is overlaid on top of GSM to provide the enhanced performance. Systems using the EDGE system may also be known as EGPRS systems.
cdmaOne	2G	1.25 MHz	CDMA	This is the brand name for the system with the standard reference IS-95, which was the first CDMA system to gain widespread use. The initial specification for the system was IS-95A, but its performance was later upgraded under IS-95B. Apart from voice, it also carries data at rates up to 14.4 kbps for IS-95A; under IS-95B, data rates of up to 115 kbps are supported.
CDMA2000 1X	2.5G	1.25 MHz	CDMA	This system supports both voice and data capabilities within a standard 1.25-MHz CDMA channel. CDMA2000 builds on cdmaOne to provide an evolution path to 3G. The system doubles the voice capacity of cdmaOne systems, and also supports high-speed data services. Peak data rates of 153 kbps are currently achievable, with figures of 307 kbps quoted for the future, and 614 kbps when two channels are used.
CDMA2000 1xEV-DO	3G	1.25 MHz	CDMA	The EV-DO stands for Evolution Data Only. This is an evolution of CDMA 2000 that is designed for data-only use, and its specification is IS-856. It provides a peak data-rate capability of over 2.45 Mbps on the forward or downlink – i.e. from the base station to the user. The aim of the system is to deliver a low cost per megabyte capability along with an always-on connection costed on the data downloaded rather than connection time.
CDMA2000 1xEV-DV	3G	1.25 MHz	CDMA	This stands for Evolution Data and Voice. It is an evolution of CDMA2000 that can simultaneously transmit voice and data. The peak data rate is 3.1 Mbps on the forward link; the reverse link is very similar to that of CDMA2000 1X and is limited to 384 kbps.
UMTS	3G	5 MHz	CDMA/TDMA	The Universal Mobile Telecommunications System uses Wideband CDMA (W-CDMA) with one 5-MHz channel for both voice and data, providing data speeds of up to 2 Mbps.
TD-SCDMA	3G	1.6 MHz	CDMA	Time Division Synchronous CDMA. A system developed in China to establish their position on the cellular telecommunications arena, it uses the same bands for transmit and receive, allowing different time slots for base stations and mobiles to communicate. Unlike other 3G systems, it uses a time division duplex (TDD) system.

that they were based on analogue technology. They separated different users in the same cell by allocating them different channels. This technique is known as Frequency Division Multiple Access (FDMA).

As demand grew, the available spectrum became progressively more congested. As a result, it quickly became obvious that less spectrum-hungry techniques would be required. As a result, the second-generation (or 2G) systems were born. These employed digital technology to provide the required levels of efficiency. The two early second-generation systems, namely GSM and US-TDMA, as well as its derivative PDC, all used a combination of FDMA and another technique whereby different users were allocated different timeslots on the same channel. This system is known as Time Division Multiple Access (TDMA). These systems offered limited data facilities. However, with revenues from voice traffic levelling off and operators seeing the opportunities for increased business as a result of data traffic, a move towards systems that could provide the performance required to make this viable was taken.

Before these high data-rate systems could be introduced to provide the high-speed data third generation or 3G systems, interim solutions were sought. These so called 2.5G systems provided higher data rates than were possible with the existing 2G systems. A system known as the General Packet Radio Service (GPRS) used with GSM provided an increase in data rate. Here, the chief change was to use a packet radio system, where individual packets of data are routed to the user, rather than using circuit-switched data, where a circuit is allocated 100 per cent to a given user. Using a packet-switched approach enables the dead periods in another call to be used productively by a user, thereby improving the efficiency of the overall system. A further data-rate improvement has been provided by another system, known as Enhanced Data rates for GSM Evolution (EDGE). Here, a different form of modulation is used to provide the data-rate increase.

Although GSM, US-TDMA and PDC use a time-division approach, another system used a different approach. Based on a spread spectrum technology, it used different codes to provide access to different users. Known as Code Division Multiple Access (CDMA), this technology was originally used on a system known as cdmaOne. This system was a 2G technology, but was upgraded through several evolutionary upgrades to provide a full 3G service.

The major cellular telecommunications systems and their capabilities are summarized in Table 1.1.

CHAPTER TWO

Radio waves and propagation

The nature of radio signals and the way in which they travel are a key elements in cellular systems. The frequencies that are used are chosen as a result of many factors, some of which are technical and others resulting from spectrum allocation issues. Nevertheless, technical issues play a key role in determining many features associated with the cellular networks and are thus of great importance. The nature of the signals, the way they travel and how they are influenced by their surroundings play a major part in determining how a network is planned.

Radio waves are a form of radiation known as electromagnetic waves. As they contain both electric and magnetic elements, it is necessary to take a look at these fields before looking at the electromagnetic wave itself.

Electric fields

Any electrically charged object, whether it has a static charge or is carrying a current, has an electric field associated with it. It is a commonly known fact that like charges repel one another and opposite charges attract. This can be demonstrated in a number of ways. For example, hair often tends to stand up after it has been brushed or combed. The brushing action generates an electrostatic charge on the hairs and, as they all have the same type of charge, they tend to repel one another and stand up. In this way it can be seen that a force is exerted between them. Examples like this are quite dramatic, and result because the voltages that are involved are very high and can typically be many kilovolts. However, even the comparatively low voltages that are found in electronic circuits exhibit the same effects, although to a much smaller degree.

The electric field radiates out from any item with an electric potential, as shown in Figure 2.1. The electrostatic potential falls away as the distance from the object is increased. Take the example of a charged sphere with a potential of 10 volts. At the surface of the sphere, the electrostatic potential is 10 V. However, as the distance from the sphere is increased, this potential starts to fall. It can be seen that it is possible to draw lines of equal potential around the sphere.

The potential falls away as the distance is increased from the sphere, and it can be shown that this occurs as the inverse of the distance – that is, doubling the distance halves the potential. The variation of potential with the distance from the sphere is shown in Figure 2.2.

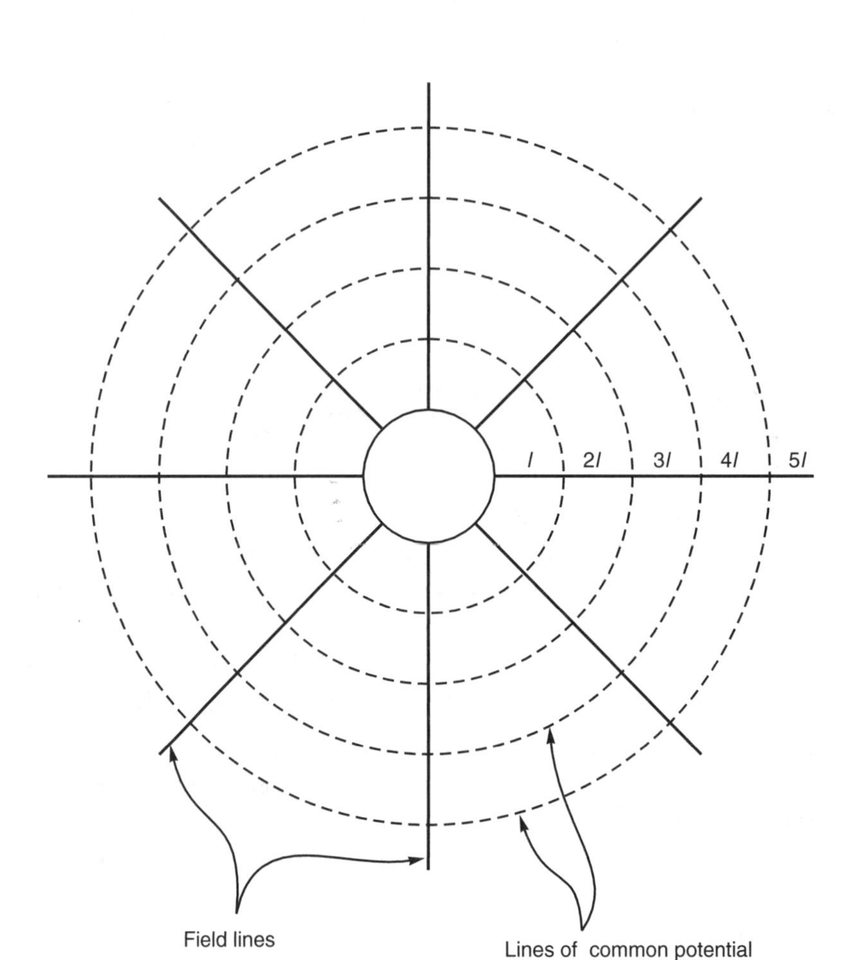

Figure 2.1 Field lines and potential lines around a charged sphere.

The electric field gives the direction and magnitude of the force on a charged object. The field intensity is the negative value of the slope in Figure 2.2. The slope of a curve plotted on a graph is the rate of change of a variable, and in this case it represents the rate of change of the potential with distance at a particular point. This is known as the *potential gradient*. It is found that the potential gradient varies as the inverse square of the distance – in other words, doubling the distance reduces the potential gradient by a factor of four.

Magnetic fields

Magnetic fields are also important. Like electric charges, magnets attract and repel one another. Analogous to the positive and negative charges, magnets have two types of pole, namely north and south. Like poles repel and dissimilar ones attract. In the case of magnets, it is also found that the magnetic field strength falls away, as the inverse square of the distance.

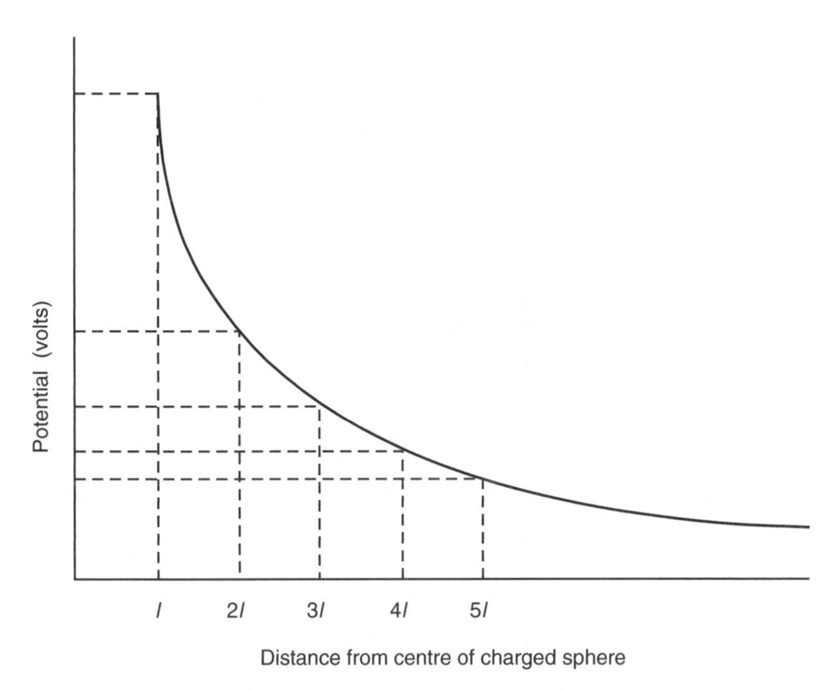

Figure 2.2 Variation of potential with distance from the charged sphere.

Whilst the first magnets to be used were permanent magnets, much later it was found that an electric current flowing in a conductor generated a magnetic field. This can be detected by the fact that a compass needle placed close to the conductor will deflect. The lines of force are in a particular direction around the wire, as shown in Figure 2.3. An easy method of determining which way they go around the conductor is to use the corkscrew rule. Imagine a right-handed corkscrew being driven into a cork in the direction of the current flow. The lines of force will be in the direction of rotation of the corkscrew.

Radio waves

As already mentioned radio signals are a form of electromagnetic wave. They consist of the same basic type of radiation as light, ultraviolet and infrared rays, differing from them in their wavelength and frequency. These waves are quite complicated in their make-up, having both electric and magnetic components that are inseparable. The planes of these fields are at right angles to one another, and to the direction of motion of the wave. These waves can be visualized as shown in Figure 2.4.

The electric field results from the voltage changes occurring in the antenna which is radiating the signal, and the magnetic changes result from the current flow. It is also found that the lines of force in the electric field run along the same axis as the antenna, but spreading out as they move away from it. This electric field is measured in terms of the change of potential over a given distance, e.g. volts per metre, and this is known as the field strength.

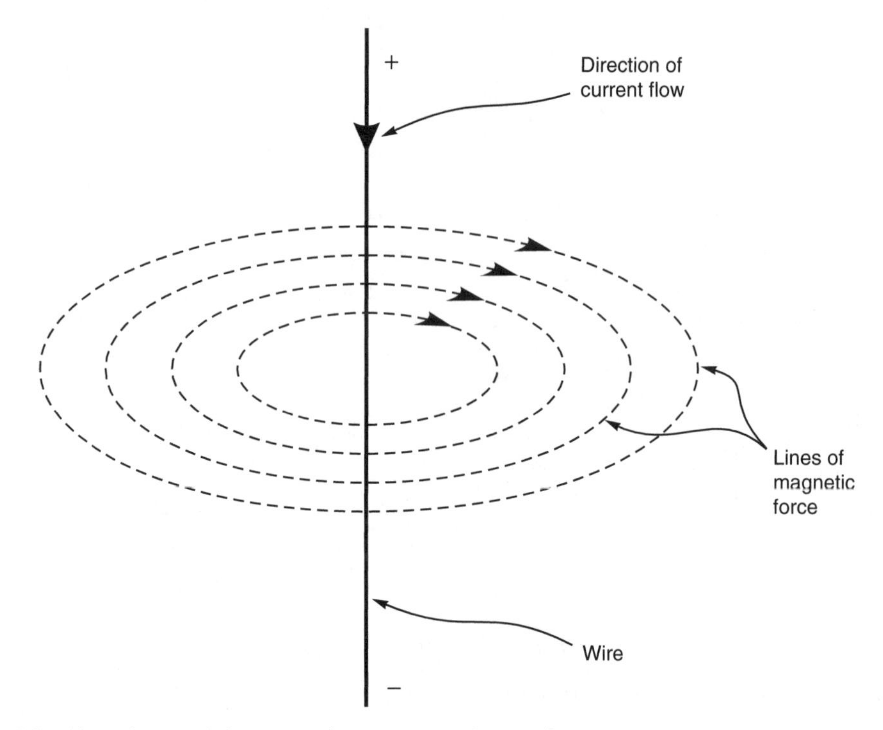

Figure 2.3 Lines of magnetic force around a current-carrying conductor.

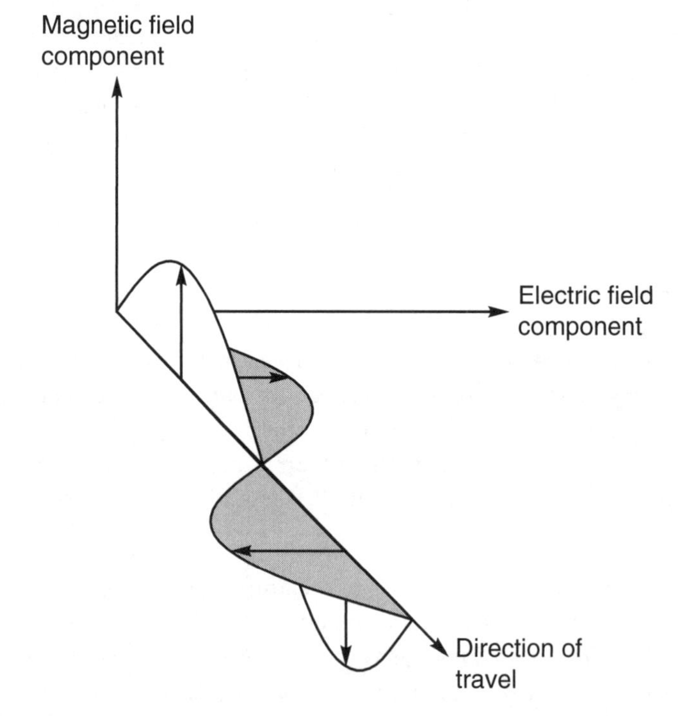

Figure 2.4 An electromagnetic wave.

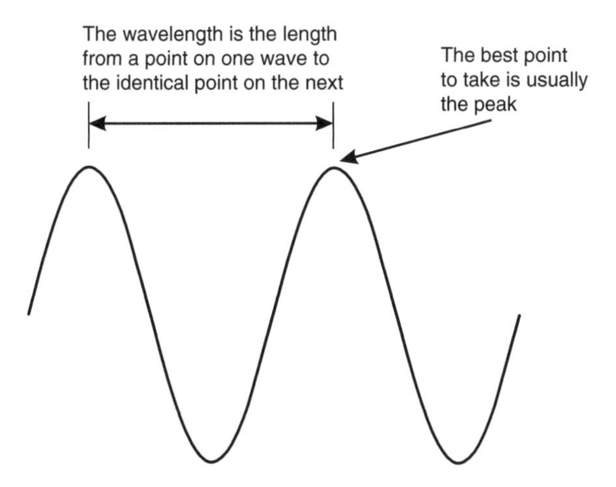

Figure 2.5 The wavelength of an electromagnetic wave.

A wave has a number of properties. The first is its wavelength. This is the distance from a point on one wave to the identical point on the next (see Figure 2.5). One of the most obvious points to choose is the peak, as this can be easily identified, although any point is acceptable.

The second property of the electromagnetic wave is its frequency. This is the number of times a particular point on the wave moves up and down in a given time (normally a second). The unit of frequency is the hertz, and it is equal to one cycle per second. This unit is named after the German scientist who discovered radio waves. The frequencies used in radio are usually very high, and accordingly the prefixes kilo, mega and giga are often seen, where 1 kHz is one thousand hertz, 1 MHz is one million hertz, and 1 GHz is one thousand million hertz (i.e. 1000 MHz). Originally the unit of frequency was not given a name, and cycles per second (c/s) was used. Some older books may show these units, together with their prefixes – kc/s; Mc/s etc. – for higher frequencies.

The third major property of the wave is its velocity. Radio waves travel at the same speed as light. For most practical purposes, the speed is taken to be 300 000 000 metres per second, although a more exact value is 299 792 500 metres per second.

Frequency-to-wavelength conversion

Many years ago, the positions of stations on the radio dial were given in terms of wavelengths – for example, a station might have had a wavelength of 1500 metres. Today stations give out their frequency, because nowadays this is far easier to measure. A frequency counter can be used to measure this very accurately and, with today's technology, their cost is relatively low. It is very easy to relate the frequency and wavelength, as they are linked by the speed of light as shown:

$$\lambda = c/f$$

where λ = wavelength in metres, f = frequency in hertz, and c = speed of radio waves (light), taken as 300 000 000 metres per second for all practical purposes.

Taking the previous example, the wavelength of 1500 metres corresponds to a frequency of 300 000 000/1500, or 200 000 hertz (200 kHz).

Radio spectrum

Electromagnetic waves have an enormously wide range of frequencies. Above the radio spectrum, other forms of radiation can be found. These include infrared radiation, light, ultraviolet radiation and a number of other forms of radiation (see Figure 2.6).

The radio portion of the electromagnetic spectrum in itself covers a large range (see Figure 2.7). At the low-frequency end of the spectrum, signals of a few kilohertz are used to communicate all over the globe at very low data rates; there are still some navigational beacons at frequencies of around 100 kHz. However, at the other end of the spectrum, signals are being generated that have frequencies of a few hundred gigahertz, although devices are now being made that have top frequencies of over 100 GHz. Between these two extremes, all the signals with which we are familiar may be found. To make it easier to refer to different areas of the radio spectrum, it is split into bands. For example, within the MF portion of the spectrum, the familiar medium-wave broadcast band may be found; above this, in the HF portion of the spectrum, are the short-wave bands. Most of the cellular telecommunications bands are around 450 MHz, 800/900 MHz, 1800/1900 MHz and 2100 MHz. This puts them all in the UHF portion of the spectrum.

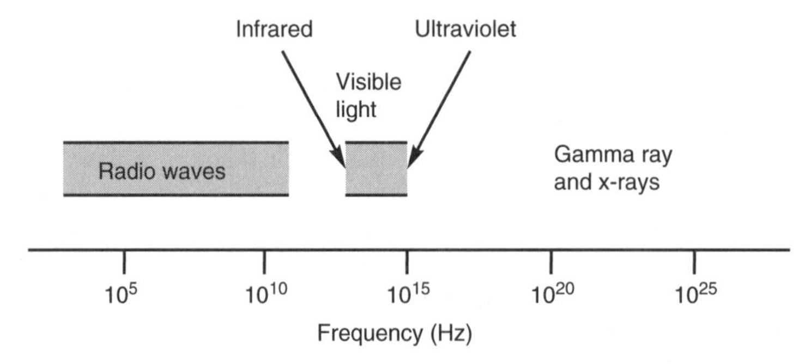

Figure 2.6 The electromagnetic wave spectrum.

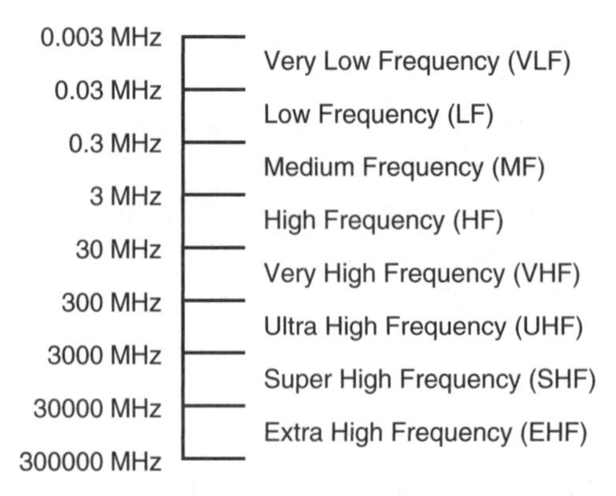

Figure 2.7 The radio spectrum.

Polarization

Apart from the frequency, another important characteristic of an electromagnetic wave is its polarization. Broadly speaking, the polarization indicates the plane in which the wave is vibrating. In view of the fact that electromagnetic waves consist of electric and magnetic components in different planes, it is necessary to define a convention. Accordingly, the polarization plane is taken to be that of the electric component.

The polarization of a radio wave can be very important, because antennas are sensitive to polarization and generally only receive or transmit a signal with a particular polarization. For most antennas it is very easy to determine the polarization; it is simply in the same plane as the elements of the antenna. Thus a vertical antenna (i.e. one with vertical elements) will receive vertically polarized signals best and, similarly, a horizontal antenna will receive horizontally polarized signals.

Vertical and horizontal are the simplest forms of polarization, and they both fall into a category known as linear polarization. However, it is also possible to use circular polarization. This has a number of benefits in areas such as satellite applications, where it helps to overcome the effects of propagation anomalies, ground reflections and the spin that occur on many satellites. Circular polarization is a little more difficult to visualize than linear polarization; however, it can be imagined by visualizing a signal propagating from an antenna that is rotating. The tip of the electric field vector will then be seen to trace out a helix or corkscrew as it travels away from the antenna. Circular polarization can be seen to be either right or left handed, dependent upon the direction of rotation as seen from the transmitter.

Another form of polarization is known as elliptical polarization. This occurs when there is a mix of linear and circular polarization, and can be visualized as before by the tip of the electric field vector tracing out an elliptically shaped corkscrew.

It can be seen that, as an antenna transmits and receives a signal with a certain polarization, the polarization of the transmitting and receiving antennas is important. This is particularly true in free space, because once a signal has been transmitted its polarization will remain the same. In order to receive the maximum signal, both transmitting and receiving antennas must be in the same plane. If for any reason their polarizations are at right angles to one another (i.e. cross-polarized) then, in theory, no signal would be received.

For terrestrial applications, it is found that once a signal has been transmitted its polarization will remain broadly the same. However, reflections from objects in the path can change the polarization. As the received signal is the sum of the direct signal plus a number of reflected signals, the overall polarization of the signal can change slightly although it remains broadly the same.

How radio signals travel

Radio signals are very similar to light waves, and behave in a comparable way. Obviously there are some differences caused by the enormous variation in frequency between the two, but in essence they are the same.

A signal may be radiated or transmitted at a certain point, and the radio waves travel outwards – much like the waves seen on a pond if a stone is dropped into it. As they move outwards they become weaker, as they have to cover a much wider area. However, they can still travel over enormous distances – light can be seen from stars many light years away. Radio waves can travel over similar distances. As distant galaxies and quasars emit radio signals these can be detected by radio telescopes, which can pick up the minute signals and then analyse them to give us further clues about what exists in the outer extremities of the universe.

While the waves on a pond become weaker as they move further outwards, the same is true of radio waves because the area on a sphere they have to cover is much greater. From simple geometry, it can be deduced that the area of the surface is proportional to the radius squared. Accordingly, the signal strength is inversely proportional to the square of the distance from the source to the receiver. This may be expressed mathematically: strength is proportional to $1/d^2$, where d is the distance from the transmitter.

These calculations are true for what is termed 'free space propagation' – that is, when the signal travels in free space and is not affected by any other objects or areas that may affect the propagation of the signal. A typical terrestrial environment is very different to this, as trees, vegetation, buildings and many other obstacles line the path of the signal, and it is found that the signal dies away at a much faster rate. Often it is closer to a rate proportional to $1/d^4$, and typically a figure of this order will be used when planning a cellular network. Some cell-phone network planners may use a factor of $1/d^{3.8}$. Whatever the exact figure used it can be seen that ranges achievable are relatively small, and therefore in order to obtain sufficient coverage for a network a large number of cell-phone masts are required.

Refraction, reflection and diffraction

With many obstacles in the path of a typical cell-phone signal travelling between a mobile phone and the base station (or *vice versa*), reflection and diffraction are important elements in understanding the signal path. Refraction may also be encountered, under some circumstances, on long paths.

In the same way that light waves can be reflected by a mirror, so radio waves can also be reflected (see Figure 2.8). When this occurs, the angle of incidence is equal to the angle of reflection for a conducting surface, as would be expected for light. When a signal is reflected there is normally some loss of the signal, either through absorption or as a result of some of the signal passing into the medium. For radio signals, surfaces such as the sea or most areas provide good reflecting surfaces, whereas desert areas are poor reflectors.

Refraction of radio waves is obviously very similar to that of light. It occurs as the wave passes through areas where the refractive index changes. For light waves this can be demonstrated by placing one end of a stick into some water, where it appears that the section of stick entering the water is bent. This occurs because the direction of the light changes as it moves

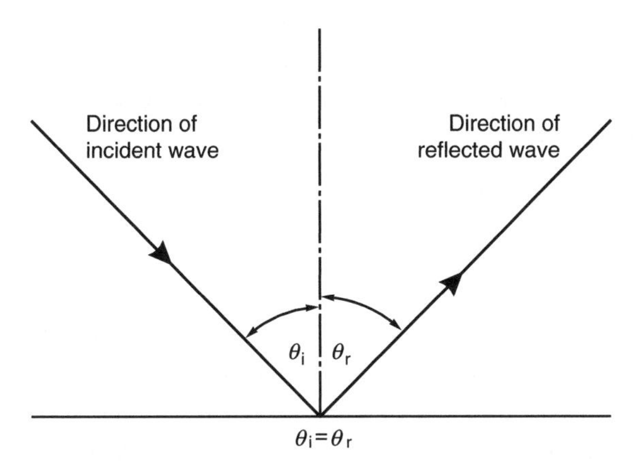

Figure 2.8 Reflection of an electromagnetic wave.

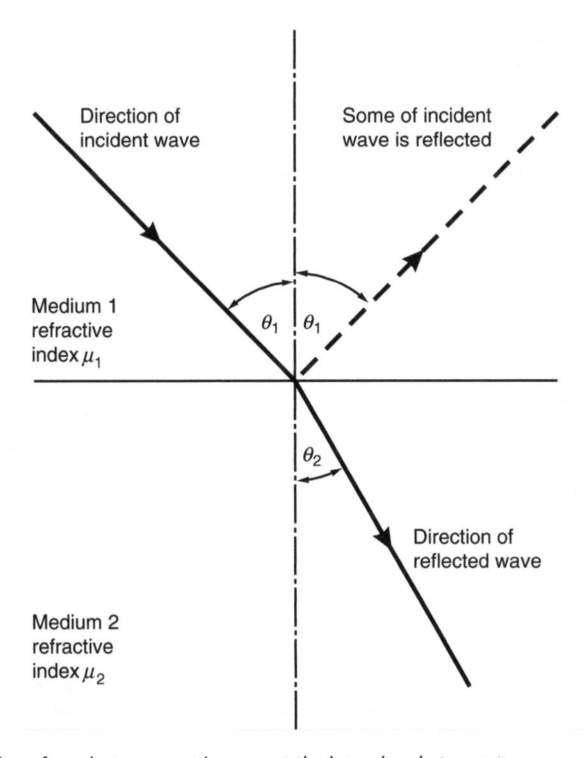

Figure 2.9 Refraction of an electromagnetic wave at the boundary between two areas of differing refractive index.

from an area of one refractive index to another. The same is true for radio waves (see Figure 2.9). In fact, the angle of incidence and the angle of refraction are linked by Snell's Law, which states that:

$$\mu_1 \sin \theta_1 = \mu_2 \sin \theta_2$$

In many cases where radio waves are travelling through the atmosphere there is a gradual change in the refractive index of the medium. This causes a steady bending of the wave rather than an immediate change in direction.

It is found that the refractive index of the areas above the Earth's surface changes slightly, with the area of higher refractive index normally being closer to the ground. This means that when signals are not obstructed by buildings and other objects, and travel over longer distances, they may be refracted by the change in refractive index. Although more pertinent to applications such as broadcasting, it is found that coverage of a station is extended beyond the line of sight by approximately one-third. As the refraction occurs in a region of the atmosphere affected by the weather, it is found that weather conditions also play a part in determining tropospheric radio propagation conditions. Under some circumstances (for example, when a stable high-pressure area is present), distances may be extended beyond their normal ranges.

Diffraction is another phenomenon that affects radio waves and light waves alike. It is found that when signals encounter an obstacle, they tend to travel around it (see Figure 2.10). The effect can be explained by Huygen's principle. This states that each point on a spherical wave front can be considered as a source of a secondary wave front. Even though there will be a shadow zone immediately behind the obstacle, the signal will diffract around the obstacle and start to fill the void, thereby enabling reception behind the obstacle even though it is not in the direct line of sight of the transmitter. It is found that diffraction is more pronounced when the obstacle approaches a 'knife edge'. A mountain ridge may provide a sufficiently sharp edge. A more rounded obstacle will not produce such a marked effect. It is also found that low-frequency signals diffract more markedly than higher-frequency signals.

Reflected signals

Signals that travel near to other objects suffer reflections from a variety of objects. One such object is the Earth itself, but others may be local buildings, or in fact anything that can reflect or partially reflect radio waves. As a result, the received signal is the sum of a variety of signals from the transmitter that have reached the receiving antenna via a number of different of paths. Each will have a slightly different path length and, as electromagnetic waves take a finite time to travel a given distance, this means that signals taking different paths arrive at the receiver at very slightly different times. Thus the signals will not reach the receiver with the same phase (see Figure 2.11).

The overall received signal is the sum of all the different signals travelling over different paths. As their phases are different, some will reinforce the strength of the overall signal whilst others will interfere with and reduce the overall strength.

Fading

Signals to and from a mobile handset will vary greatly in strength as the user moves from place to place. The signal variations will be very large, and often occur over very small distances. This means that they are not primarily due to the distance changing, although naturally as the mobile moves further away from the base station the path length increases and the signal falls.

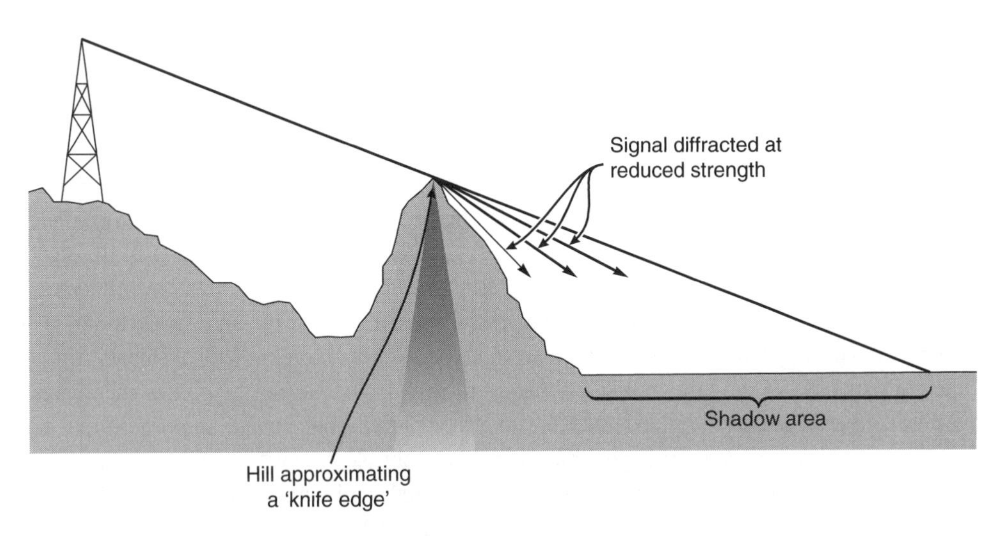

Figure 2.10 Diffraction of a radio signal around an obstacle.

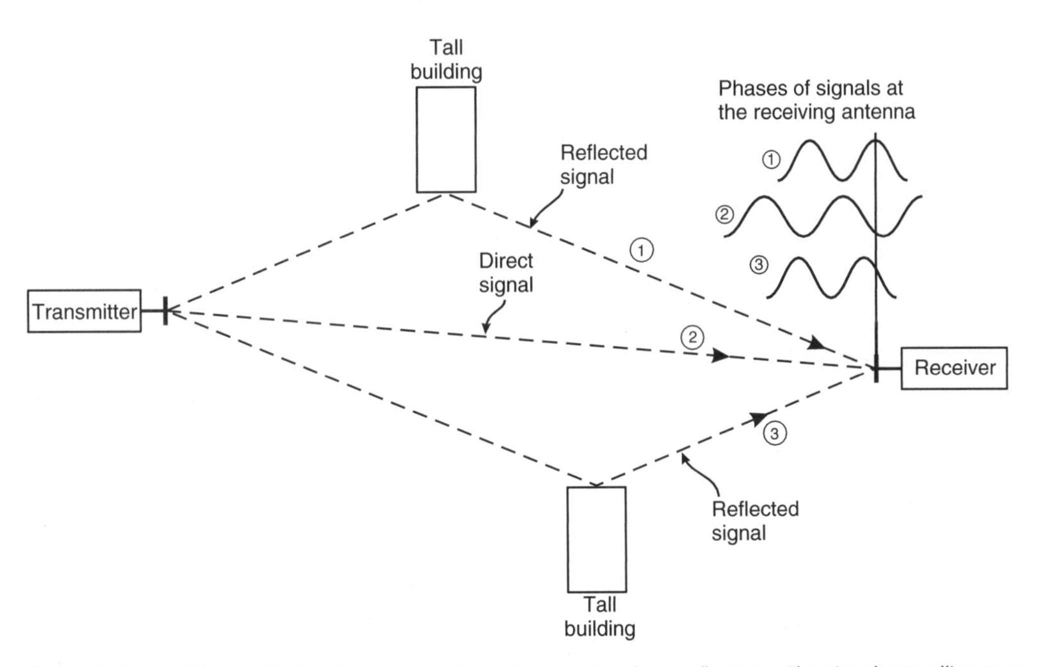

Figure 2.11 Multiple paths lengths of a received signal arising from reflections. The signals travelling over these paths arrive with different phases.

Fading may be categorized into two types, namely slow fading and fast fading, their names giving a description of the noticed effect on the signal strength.

Slow fading may be caused by the mobile phone passing behind an obstruction, such as a building, that masks the signal to and from the base station. It is found that as the phone passes behind the obstruction, the signal strength will fall over the space of several metres, dependent upon the nature of the obstruction and distance from it. When the phone is travelling or moving, this gives a fade that is relatively slow when compared to fast fading, which is explained next.

The other form of fading is, not surprisingly, called fast fading. It arises from the fact that the signal reaches the receiver via several paths. The direct or line-of-sight (LOS) path is the most obvious, but reflections also make up part of the signal reaching the receiver, resulting in what is termed 'multipath' propagation. The total signal picked up by the receiver is a combination of the signal received via the LOS path as well as those from reflections. These signals will all have different phases because they have travelled over different paths and have taken different times to arrive at the receiver. Accordingly, the overall received signal is the sum of all the individual signals arriving via different paths. By moving the mobile phone, even by a small amount, the phases will change, and with this so can the overall signal level. To gain a view of the small changes in position that can give rise to a significant change, it is possible to look at a simple case where just two signals are received, one via the direct LOS path and another from a reflected path. Assuming a frequency of 2 GHz, it can be seen that the wavelength is $c/f = 3 \times 10^8/2 \times 10^9 = 0.15$ m. To move from a signal being in phase to a signal being out of phase is equivalent to increasing the path length by half a wavelength, or 0.075 m (7.5 cm). The situation is naturally not as simple as this, because the overall signal consists of many different signals, but is gives an idea of the distance needed to be moved to change from an in-phase to an out of phase situation. When moving even relatively slowly, fading as a result of this phenomenon will occur relatively swiftly, hence the name – fast fading.

Intersymbol interference

A further problem that can be caused by reflections is known as InterSymbol Interference (ISI). It occurs in systems that are transmitting digital information. Typically the data rates being sent may be of the order of several kilobits per second, as in the case of a second-generation system. It occurs when signals that have been reflected by distant objects are received. As the path length has been increased by a large degree, it is possible that the time delay can be such that the receiver may be receiving a signal via the direct path that may be one bit of data, whereas the reflected signal may be delayed to a sufficient extent that it is carrying the previous bit of data.

For example, a data rate may be 50 kbps. The time that it takes for one bit of data to be sent is 1/50 000 s, or 20 μs. During this time, a radio signal would travel 6 km – in other words, the reflected signal path would need to be 6 km longer than that of the direct signal. This additional path length could occur if the signal were reflected by a building or other object just 3 km behind the mobile phone. Many of today's buildings, with their lined windows, form very good reflective surfaces, and often reflections can be very strong. Also with increasing data rates, this problem could be more acute, although there are methods of overcoming the problem.

Attenuation by the atmosphere

It is possible for the atmosphere to introduce path loss beyond that normally encountered by the free-space spreading effect. For transmissions in the UHF section of the spectrum, atmospheric conditions such as rain and fog have little effect on the signals. However, as the frequency increases the atmosphere has a much greater effect on the level of attenuation in the signal path, and at certain frequencies the loss that is introduced has to be considered.

However, as the frequencies rise above about 3 GHz the loss can introduce an additional degree of variation into the path. As may be expected, the loss is dependent upon the amount of rain and also the size of the droplets. As a rough guide, very heavy rain may introduce an additional loss of about 1 dB per kilometre at around 5 GHz, and more at higher frequencies. The loss occurs for two reasons. The first is absorption by the rain droplets, with the level of actual attenuation being dependent upon the droplet size. The second occurs as a result of the signal being scattered, and although the power is not lost, not all of it travels in the original direction it was intended. In this way, the antenna gain is effectively reduced.

At frequencies well above 10 GHz, attenuation arising from the gases in the air may be evident – in particular, water vapour and oxygen. This arises by virtue of the permanent electric dipole moment of the water vapour and the permanent magnetic dipole moment of the oxygen molecule. There are peaks for the components, with an oxygen peak around 60 GHz giving rise to an attenuation of around 15 dB per kilometre, and a lower peak at just over 100 GHz giving rise to attenuation of just under 2 dB per kilometre. Water vapour losses rise steadily with frequency but peak just below 200 GHz, introducing a loss of nearly 40 dB per kilometre. For the current frequencies in use by cellular operators these effects are likely to have little impact, although if new bands with much higher frequencies are introduced, this may change.

Coverage and network planning

When planning and maintaining a network, it is necessary to be able to determine what the coverage of the different base stations located at different points will be. This must be achieved in the planning stage, and not left until the network is being deployed. Similarly, it is necessary to be able to plan the coverage of any new base stations being introduced into an existing network before they are installed. Only by doing this is it possible to achieve the optimum coverage with the most efficient use of the base stations.

To achieve this, computer coverage prediction programmes are used. These use maps in a digital format, along with propagation algorithms and details of the base station, to predict the actual coverage. In this way the network can be simulated before it is deployed, saving considerable amounts of money, time and effort, and enabling a far more efficient network to be deployed from the start.

Modulation

Radio signals can be used to carry information. The information, which may be audio, data or other forms, is used to modify (modulate) a single frequency known as the carrier. The information superimposed onto the carrier forms a radio signal which is transmitted to the receiver. Here, the information is removed from the radio signal and reconstituted in its original format in a process known as demodulation. It is worth noting at this stage that the carrier itself does not convey any information.

There are many different varieties of modulation but they all fall into three basic categories, namely amplitude modulation, frequency modulation and phase modulation, although frequency and phase modulation are essentially the same. Each type has its own advantages and disadvantages. A review of all three basic types will be undertaken, although a much greater focus will be placed on those types used within phone systems. By reviewing all the techniques, a greater understanding of the advantages and disadvantages can be gained.

Radio carrier

The basis of any radio signal or transmission is the carrier. This consists of an alternating waveform like that shown in Figure 3.1. This is generated in the transmitter, and if it is radiated in this form it carries no information – it appears at the receiver as a constant signal.

Amplitude modulation

Possibly the most obvious method of modulating a carrier is to change its amplitude in line with the modulating signal.

The simplest form of amplitude modulation is to employ a system known as 'on–off keying' (OOK), where the carrier is simply turned on and off. This is a very elementary form of digital modulation and was the method used to carry Morse transmissions, which were widely used especially in the early days of 'wireless'. Here, the length of the on and off periods defined the different characters.

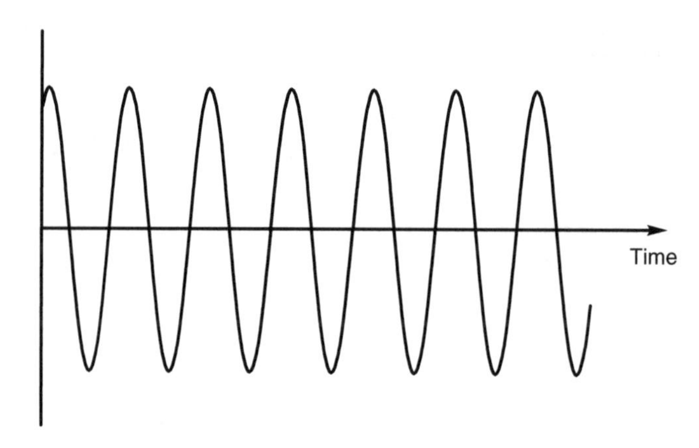

Figure 3.1 An alternating waveform.

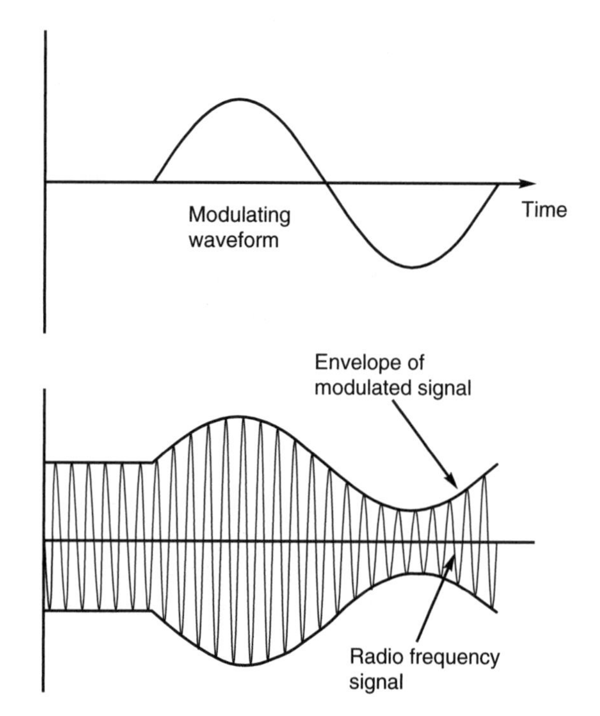

Figure 3.2 An amplitude modulated signal.

More generally, the amplitude of the overall signal is varied in line with the incoming audio or other modulating signal, as shown in Figure 3.2. Here, the envelope of the carrier can be seen to change in line with the modulating signal. This is known as Amplitude Modulation (AM).

The demodulation process for AM where the radio frequency signal is converted into an audio frequency signal is very simple. It only requires a simple diode detector circuit like that shown in Figure 3.3. In this circuit the diode rectifies the signal, only allowing the one-half of the

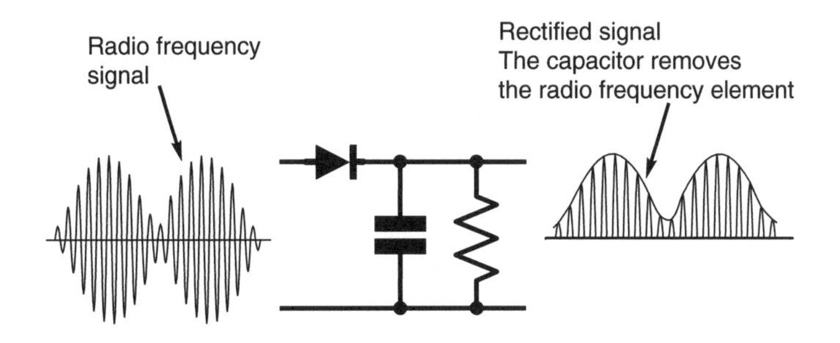

Figure 3.3 A simple diode detector circuit.

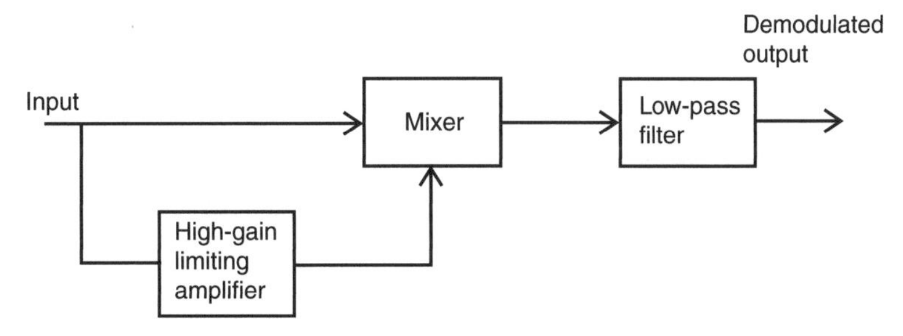

Figure 3.4 Synchronous AM demodulation.

alternating radio frequency waveform through. A capacitor is used as a simple low-pass filter to remove the radio-frequency parts of the signal, leaving the audio waveform. This can be fed into an amplifier, after which it can be used to drive a loudspeaker. This form of demodulator is very cheap and easy to implement, and is still widely used in many AM receivers today.

The signal may also be demodulated more efficiently using a system known as synchronous detection (Figure 3.4). Here, the signal is mixed with a locally generated signal with the same frequency and phase as the carrier. In this way the signal is converted down to the baseband frequency. This system has the advantage of a more linear demodulation characteristic than the diode detector, and it is more resilient to various forms of distortion. There are various methods of generating the mix signal. One of the easiest is to take a feed from the signal being received and pass it through a very high-gain amplifier. This removes any modulation, leaving just the carrier with exactly the required frequency and phase. This can be mixed with the incoming signal and the result filtered to recover the original audio.

AM has the advantage of simplicity, but it is not the most efficient mode to use – both in terms of the amount of spectrum it takes up and the usage of the power. For this reason, it is rarely used for communications purposes. Its only major communications use is for VHF aircraft communications. However, it is still widely used on the long, medium, and short wave bands for broadcasting because its simplicity enables the cost of radio receivers to be kept to a minimum.

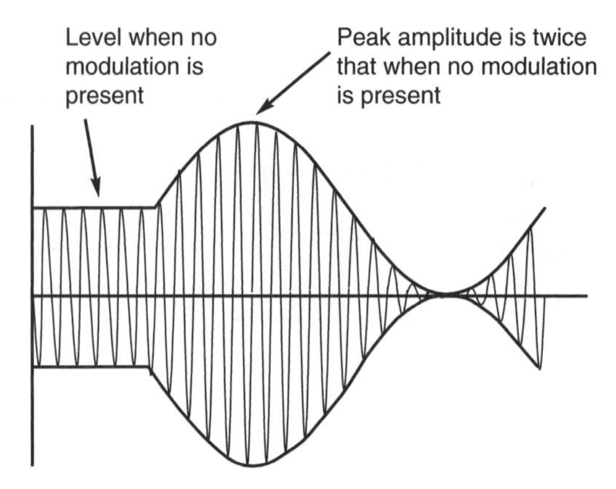

Figure 3.5 Fully modulated signal.

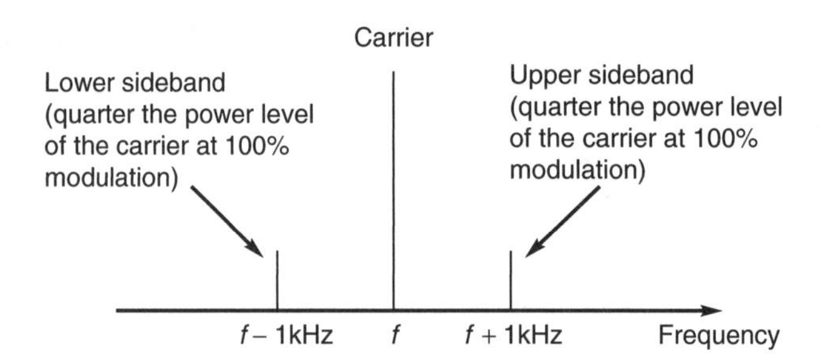

Figure 3.6 Spectrum of a signal modulated with a 1-kHz tone.

To find out why it is inefficient, it is necessary to look at a little theory behind the operation of AM. When a radio-frequency signal is modulated by an audio signal, the envelope will vary. The level of modulation can be increased to a level where the envelope falls to zero and then rises to twice the unmodulated level. Any increase above this will cause distortion because the envelope cannot fall below zero. As this is the maximum amount of modulation possible, it is called 100 per cent modulation (Figure 3.5).

Even with 100 per cent modulation, the utilization of power is very poor. When the carrier is modulated, sidebands appear at either side of the carrier in its frequency spectrum. Each sideband contains the information about the audio modulation. To look at how the signal is made up and the relative powers, take the simplified case where the 1-kHz tone is modulating the carrier. In this case, two signals will be found: 1 kHz either side of the main carrier, as shown in Figure 3.6. When the carrier is fully modulated (i.e. 100 per cent), the amplitude of the modulation is equal to half that of the main carrier – that is, the sum of the powers of the sidebands is equal to half that of the carrier. This means that each sideband is just a quarter of the total power. In other words, for a transmitter with a 100-watt carrier, the total sideband power will be 50 W and each individual sideband will be 25 W. During the modulation process the carrier power remains constant. It is only needed as a reference during the demodulation process. This means that the sideband power is the useful section of the signal, and this corresponds to (50/150) × 100 per cent, or only 33 per cent of the total power transmitted.

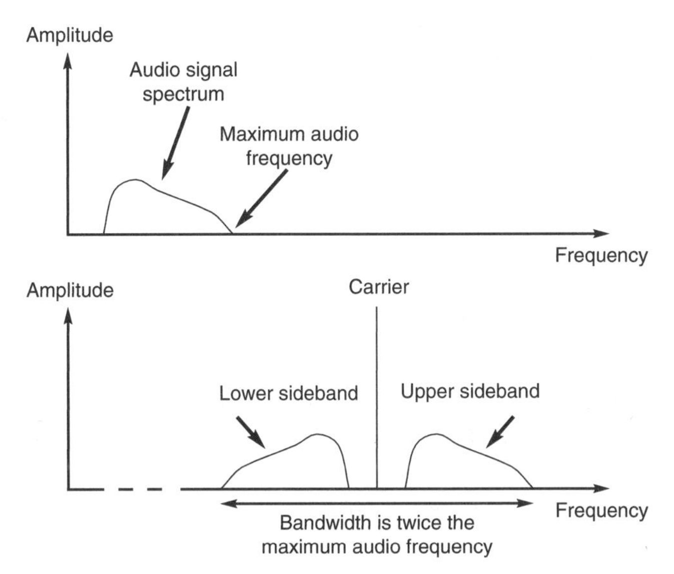

Figure 3.7 Spectrum of a signal modulated with speech or music.

Not only is AM wasteful in terms of power; it is also not very efficient in its use of spectrum. If the 1-kHz tone is replaced by a typical audio signal made up of a variety of sounds with different frequencies, then each frequency will be present in each sideband (Figure 3.7). Accordingly, the sidebands spread out either side of the carrier as shown and the total bandwidth used is equal to twice the top frequency that is transmitted. In the crowded conditions found on many of the short wave bands today this is a waste of space, and other modes of transmission that take up less space are often used.

To overcome the disadvantages of AM, a derivative known as single sideband (SSB) is often used. By removing or reducing the carrier and removing one sideband, the bandwidth can be halved and the efficiency improved. The carrier can be introduced by the receiver for demodulation.

Neither AM in its basic form nor SSB is used for mobile phone applications, although in some applications AM combined with phase modulation is used.

Modulation index

It is often necessary to define the level of modulation that is applied to a signal. A factor or index known as the modulation index is used for this. When expressed as a percentage, it is the same as the depth of modulation. In other words, it can be expressed as:

$$M = \frac{\text{RMS value of modulating signal}}{\text{RMS value of unmodulated signal}}.$$

The value of the modulation index must not be allowed to exceed 1 (i.e. 100 per cent in terms of the depth of modulation), otherwise the envelope becomes distorted and the signal will spread out either side of the wanted channel, causing interference to other users.

Frequency modulation

While AM is the simplest form of modulation to envisage, it is also possible to vary the frequency of the signal to give frequency modulation (FM). It can be seen from Figure 3.8 that the frequency of the signal varies as the voltage of the modulating signal changes.

The amount by which the signal frequency varies is very important. This is known as the deviation, and is normally quoted in kilohertz. As an example, the signal may have a deviation of ±3 kHz. In this case, the carrier is made to move up and down by 3 kHz.

FM is used for a number of reasons. One particular advantage is its resilience to signal-level variations and general interference. The modulation is carried only as variations in frequency, and this means that any signal-level variations will not affect the audio output provided that the signal is of a sufficient level. As a result, this makes FM ideal for mobile or portable applications where signal levels vary considerably. The other advantage of FM is its resilience to noise and interference when deviations much greater than the highest modulating frequency are used. It is for this reason that FM is used for high-quality broadcast transmissions where deviations of ±75 kHz are typically used to provide a high level of interference rejection. In view of these advantages, FM was chosen for use in the first-generation analogue mobile phone systems.

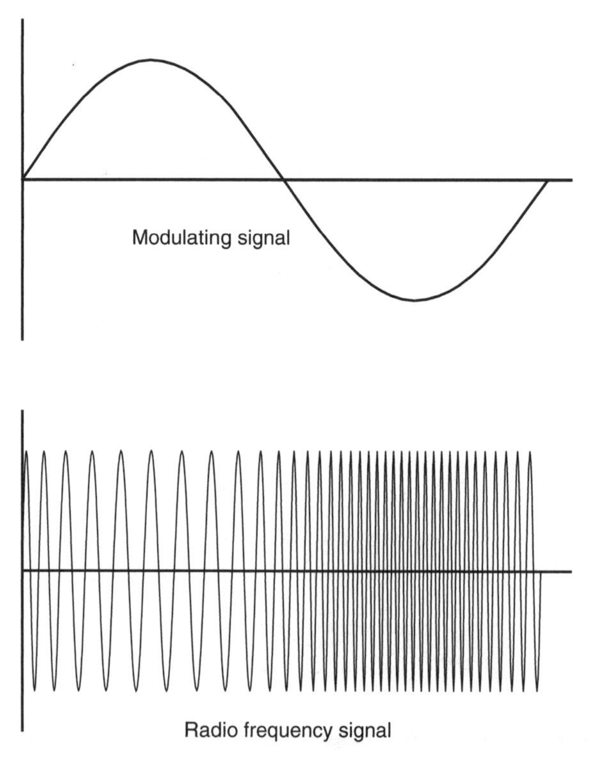

Figure 3.8 A frequency modulated signal.

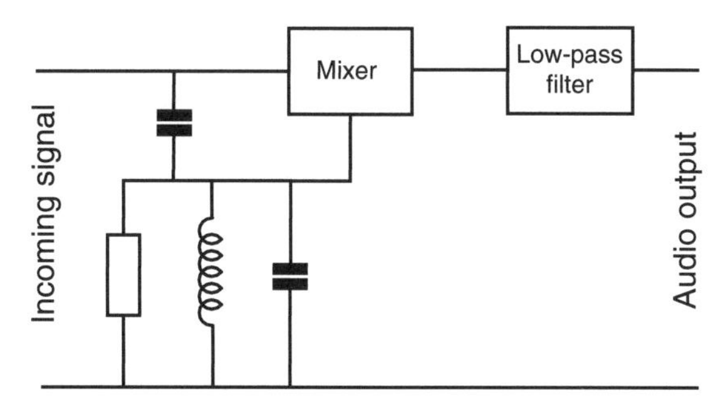

Figure 3.9 Block diagram of an FM quadrature detector.

To demodulate an FM signal, it is necessary to convert the frequency variations into voltage variations. This is slightly more complicated than demodulating AM, but it is still relatively simple to achieve. Rather than just detecting the amplitude level using a diode, a tuned circuit has to be incorporated so that a different output voltage level is given as the signal changes its frequency. There is a variety of methods used to achieve this, but one popular approach is to use a system known as a quadrature detector. It is widely used in integrated circuits, and provides a good level of linearity. It has the advantages that it requires a simple tuned circuit and it is also very easy to implement in a form that is applicable to integrated circuits.

The basic format of the quadrature detector is shown in Figure 3.9. It can be seen that the signal is split into two components. One of these passes through a network that provides a basic 90° phase shift, plus an element of phase shift dependent upon the deviation. The original signal and the phase-shifted signal are then passed into a multiplier or mixer. The mixer output is dependent upon the phase difference between the two signals, i.e. it acts as a phase detector and produces a voltage output that is proportional to the phase difference and hence to the level of deviation of the signal.

Modulation index and deviation ratio

In many instances a figure known as the modulation index is of value and is used in other calculations. The modulation index is the ratio of the frequency deviation to the modulating frequency, and will therefore vary according to the frequency that is modulating the transmitted carrier and the amount of deviation:

$$M = \frac{\text{Frequency deviation}}{\text{Modulation frequency}}.$$

However, when designing a system it is important to know the maximum permissible values. This is given by the deviation ratio, and is obtained by inserting the maximum values into the formula for the modulation index:

$$D = \frac{\text{Maximum frequency deviation}}{\text{Maximum modulation frequency}}.$$

Sidebands

Any signal that is modulated produces sidebands. In the case of an amplitude modulated signal they are easy to determine, but for frequency modulation the situation is not quite as straightforward. They are dependent upon not only the deviation, but also the level of deviation – i.e. the modulation index M. The total spectrum is an infinite series of discrete spectral components, expressed by the complex formula:

$$\begin{aligned}\text{Spectrum components} = Vc\{&J_0(M)\cos\omega_c t \\ &+J_1(M)[\cos(\omega_c+\omega_m)t-\cos(\omega_c-\omega_m)t] \\ &+J_2(M)[\cos(\omega_c+2\omega_m)t-\cos(\omega_c-2\omega_m)t] \\ &+J_3(M)[\cos(\omega_c+3\omega_m)t-\cos(\omega_c-3\omega_m)t] \\ &+\ldots\}.\end{aligned}$$

In this relationship, $J_n(M)$ are Bessel functions of the first kind, ω_c is the angular frequency of the carrier and is equal to $2\pi f$, and ω_m is the angular frequency of the modulating signal. Vc is the voltage of the carrier.

It can be seen that the total spectrum consists of the carrier plus an infinite number of sidebands spreading out on either side of the carrier at integral frequencies of the modulating frequency. The relative levels of the sidebands can be read from a table of Bessel functions, or calculated using a suitable computer program. Figure 3.10 shows the relative levels to give an indication of the way in which the levels of the various sidebands change with different values of modulation index.

It can be gathered that for small levels of deviation (that is, what is termed narrowband FM) the signal consists of the carrier and the two sidebands spaced at the modulation frequency either side

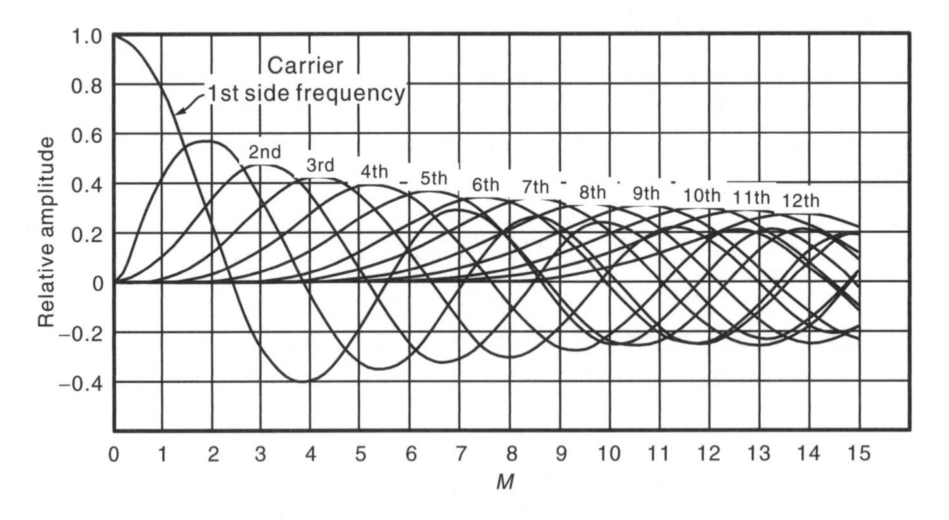

Figure 3.10 The relative amplitudes of the carrier and the first 10 side frequency components of a frequency modulated signal for different values of modulation index.

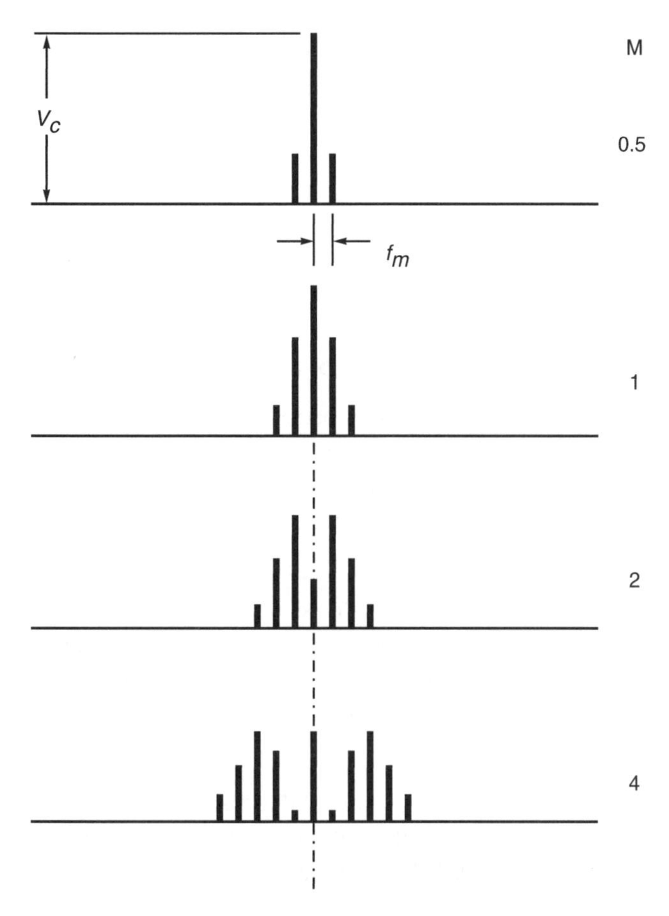

Figure 3.11 Spectra of frequency-modulated signals with various values of modulation index for a constant modulation frequency. It can be seen that for small values of the modulation index M (e.g. M = 0.5), the signal appears to consist of the carrier and two sidebands. As the modulation index increases, the number of sidebands increases and the level of the carrier can be seen to decrease for these values.

of the carrier. The spectrum appears the same as that of an AM signal. The major difference is that the lower sideband is out of phase by 180°.

As the modulation index increases, other sidebands at twice the modulation frequency start to appear (Figure 3.11). As the index is increased, further sidebands can also be seen. It is also found that the relative levels of these sidebands change, some rising in level and others falling as the modulation index varies.

Bandwidth

It is clearly not acceptable to have a signal that occupies an infinite bandwidth. Fortunately, for low levels of modulation index all but the first two sidebands may be ignored. However, as the modulation index increases the sidebands further out increase in level, and it is often necessary to apply filtering to the signal. This should not introduce any undue distortion. To achieve this it is normally necessary to allow a bandwidth equal to twice the maximum frequency of deviation plus the maximum modulation frequency. In other words, for a VHF FM broadcast station with a deviation of ±75 kHz and a maximum modulation frequency of 15 kHz, this must be (2 × 75) + 15 kHz, i.e. 175 kHz. In view of this a total of 200 kHz is usually allowed, enabling stations to have a small guard band and their centre frequencies on integral numbers of 100 kHz.

Improvement in signal-to-noise ratio

It has already been mentioned that FM can give a better signal-to-noise ratio than AM when wide bandwidths are used. The amplitude noise can be removed by limiting the signal. In fact, the greater the deviation, the better the noise performance. When comparing an AM signal with an FM signal, an improvement equal to $3D^2$ is obtained where D is the deviation ratio. This is true for high values of D – i.e. wideband FM.

An additional perceived improvement in signal-to-noise ratio can be achieved if the audio signal is pre-emphasized. To achieve this, the lower-level high-frequency sounds are amplified to a greater degree than the lower-frequency sounds before they are transmitted. Once at the receiver, the signals are passed through a network with the opposite effect to restore a flat frequency response.

To achieve the pre-emphasis, the signal may be passed through a capacitor–resistor (CR) network. At frequencies above the cut-off frequency, the signal increases in level by 6 dB per octave. Similarly, at the receiver the response falls by the same amount.

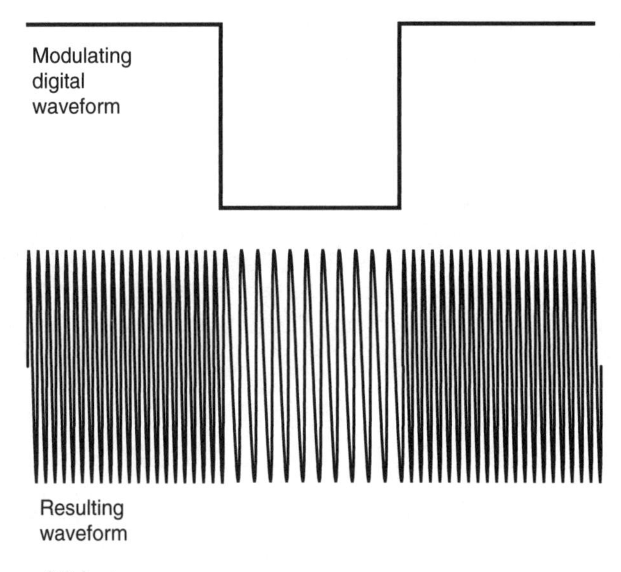

Figure 3.12 Frequency shift keying.

Frequency shift keying

Many signals employ a system called frequency shift keying (FSK) to carry digital data (Figure 3.12). Here, the frequency of the signal is changed from one frequency to another, one frequency counting as the digital 1 (mark) and the other as a digital 0 (space). By changing the frequency of the signal between these two it is possible to send data over the radio.

There are two methods that can be employed to generate the two different frequencies needed for carrying the information. The first and most obvious is to change the frequency of the carrier. Another method is to frequency-modulate the carrier with audio tones that change in frequency, in a scheme known as Audio Frequency Shift Keying (AFSK). This second method can be of advantage when tuning accuracy is an issue.

Phase modulation

Another form of modulation that is widely used, especially for data transmissions, is Phase Modulation (PM). As phase and frequency are inextricably linked (frequency being the rate of change of phase), both forms of modulation are often referred to by the common term 'angle modulation'.

To explain how phase modulation works, it is first necessary to give an explanation of phase. A radio signal consists of an oscillating carrier in the form of a sine wave. The amplitude follows this curve, moving positive and then negative, and returning to the start point after one complete cycle. This can also be represented by the movement of a point around a circle, the phase at any given point being the angle between the start point and the point on the waveform as shown in Figure 3.13.

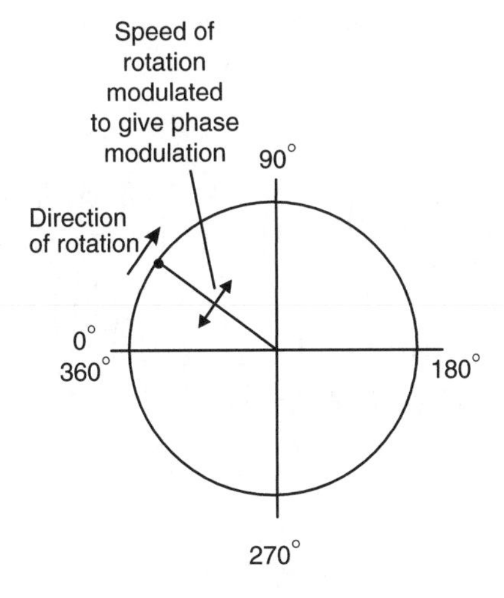

Figure 3.13 Phase modulation.

Modulating the phase of the signal changes the phase from what it would have been if no modulation were applied. In other words, the speed of rotation around the circle is modulated about the mean value. To achieve this it is necessary to change the frequency of the signal for a short time. In other words, when phase modulation is applied to a signal there are frequency changes and *vice versa*. Phase and frequency are inseparably linked, as phase is the integral of frequency. Frequency modulation can be changed to phase modulation by simply adding a CR network to the modulating signal that integrates the modulating signal. As such, the information regarding sidebands, bandwidth and the like also holds true for phase modulation as it does for frequency modulation, bearing in mind their relationship.

Phase shift keying

Phase modulation may be used for the transmission of data. Frequency shift keying is robust, and has no ambiguities because one tone is higher than the other. However, phase shift keying (PSK) has many advantages in terms of efficient use of bandwidth and is the form of modulation chosen for many cellular telecommunications applications.

The basic form of phase shift keying is known as Binary Phase Shift Keying (BPSK) or, occasionally, Phase Reversal Keying (PRK). A digital signal alternating between +1 and −1 (or 1 and 0) will create phase reversals – i.e. 180° phase shifts – as the data shifts state (Figure 3.14).

The problem with phase shift keying is that the receiver cannot know the exact phase of the transmitted signal, to determine whether it is in a mark or space condition. This would not be possible even if the transmitter and receiver clocks were accurately linked, because the path length would determine the exact phase of the received signal. To overcome this problem, PSK systems use a differential method for encoding the data onto the carrier. This is accomplished by, for example, making a change in phase equal to a 1 and no phase change equal to a 0. Further improvements can be made upon this basic system, and a number of other types of phase shift keying have been developed. One simple improvement can be made by making a change in phase of 90° in one direction for a 1, and 90° the other way for a 0. This retains the 180° phase reversal between the 1 and 0 states, but gives a distinct change for a 0. In a basic system

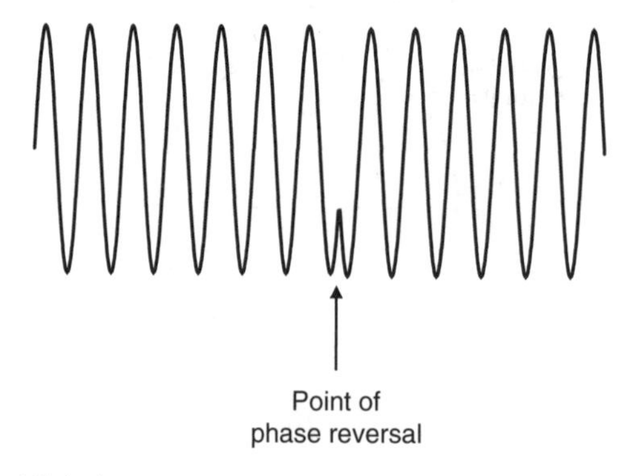

Figure 3.14 Binary phase shift keying.

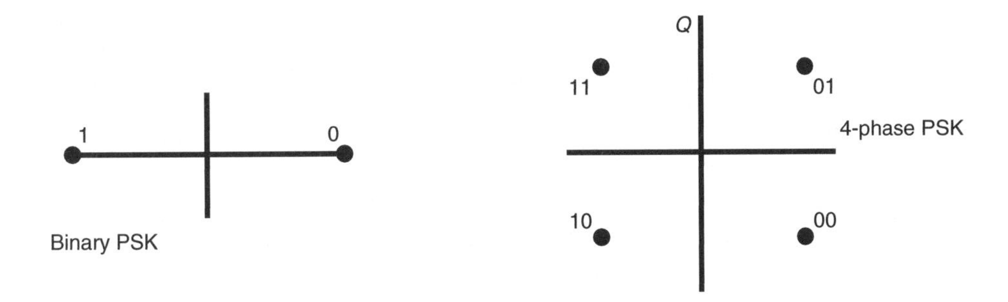

Figure 3.15 Phasor constellations for BPSK and QPSK.

not using this process it may be possible to lose synchronization if a long series of zeros is sent. This is because the phase will not change state for this occurrence.

There are many variations on the basic idea of phase shift keying. Each has its own advantages and disadvantages, enabling system designers to choose the one most applicable for any given circumstances. Other common forms include Quadrature Phase Shift Keying (QPSK), where four phase states are used, each at 90° to the other; 8-PSK, where there are eight states and so forth.

It is often convenient to represent a phase shift keyed signal, and sometimes other types of signal, using a phasor or constellation diagram (see Figure 3.15). Using this scheme, the phase of the signal is represented by the angle around the circle, and the amplitude by the distance from the origin or centre of the circle. In this way the signal can be resolved into quadrature components representing the sine or I for In-phase component, and the cosine for the quadrature component. Most phase-shift-keyed systems use a constant amplitude, and therefore points appear on one circle with a constant amplitude and the changes in state being represented by movement around the circle. For binary shift keying using phase reversals, the two points appear at opposite points on the circle. Other forms of phase shift keying may use different points on the circle, and there can be more points on the circle.

When plotted using test equipment, errors may be seen from the ideal positions on the phase diagram. These errors may appear as the result of inaccuracies in the modulator and transmission and reception equipment, or as noise that enters the system. It can be imagined that if the position of the real measurement when compared to the ideal position becomes too large, then data errors will appear because the receiving demodulator is unable correctly to detect the intended position of the point on the circle.

Using a constellation view of the signal enables quick fault-finding in a system. If the problem is related to phase, the constellation will spread around the circle. If the problem is related to magnitude, the constellation will spread off the circle, either towards or away from the origin. These graphical techniques assist in isolating problems much faster than when using other methods.

QPSK is used for the forward link from the base station to the mobile in the IS-95 cellular system, and uses the absolute phase position to represent the symbols. There are four phase decision points, and when transitioning from one state to another it is possible to pass through the circle's origin, indicating minimum magnitude.

On the reverse link from mobile to base station, Offset-Quadrature Phase Shift Keying (O-QPSK) is used to prevent transitions through the origin. Consider the components that make up any particular vector on the constellation diagram as X and Y components. Normally, both of these components would transition simultaneously, causing the vector to move through the origin. In O-QPSK one component is delayed, so the vector will move down first and then over, thus avoiding moving through the origin, and simplifying the radio's design. A constellation diagram will show the accuracy of the modulation.

Minimum shift keying

It is found that binary data consisting of sharp transitions between '1' and '0' states and *vice versa* potentially create signals that have sidebands extending out a long way from the carrier, and this is not ideal from many aspects. This can be overcome in part by filtering the signal, but the transitions in the data become progressively less sharp as the level of filtering is increased and the bandwidth is reduced. To overcome this, a form of modulation known as Gaussian-filtered Minimum Shift Keying (GMSK) is widely used – for example, it has been adopted for the GSM standard for mobile telecommunications. It is derived from a modulation scheme known as Minimum Shift Keying (MSK), which is what is known as a continuous-phase scheme. Here, there are no phase discontinuities because the frequency changes occur at the carrier zero crossing points.

To illustrate this, take the example shown in Figure 3.16. Here, it can be seen that the modulating data signal changes the frequency of the signal and there are no phase discontinuities. This arises as a result of the unique factor of MSK that the frequency difference between the logical 1 and logical 0 states is always equal to half the data rate. This can be expressed in terms of the modulation index, and is always equal to 0.5.

Whilst this method appears to be fine, in fact the bandwidth occupied by an MSK signal is too wide for many systems, where a maximum bandwidth equal to the data rate is required.

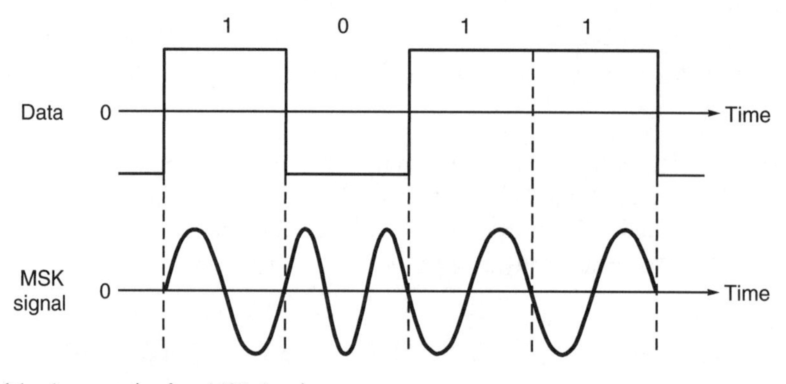

Figure 3.16 An example of an MSK signal.

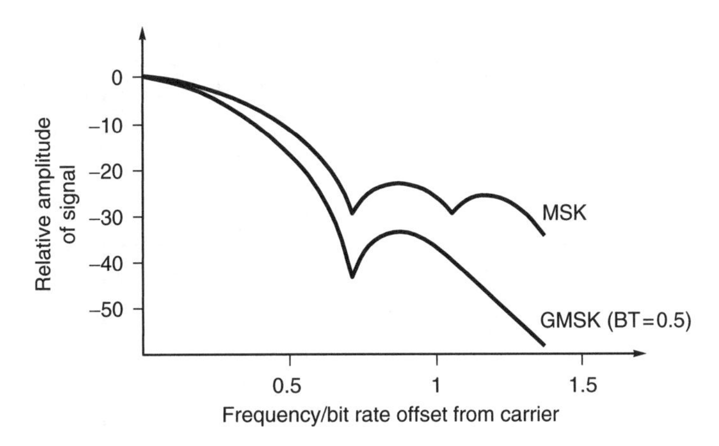

Figure 3.17 Graph of the spectral density for MSK and GMSK signals.

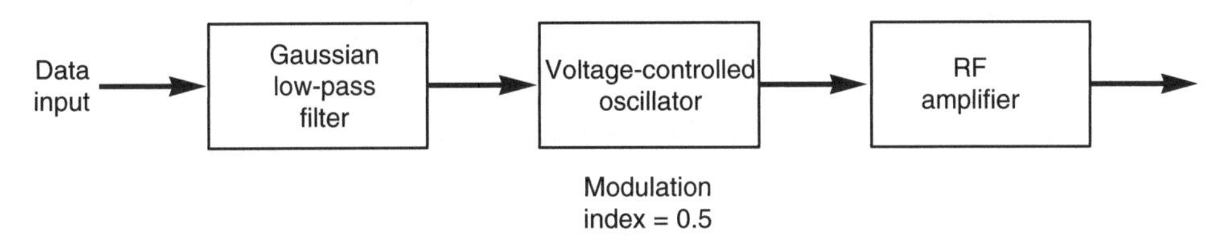

Figure 3.18 Generating GMSK using a Gaussian filter and a frequency modulator with the modulation index set to 0.5.

A plot of the spectrum of an MSK signal shows sidebands extending well beyond a bandwidth equal to the data rate (Figure 3.17). This can be reduced by passing the modulating signal through a low-pass filter prior to applying it to the carrier. The requirements for the filter are that it should have a sharp cut-off and a narrow bandwidth, and its impulse response should show no overshoot. The ideal filter is known as a Gaussian filter, which has a Gaussian-shaped response to an impulse and no ringing.

There are two main ways in which GMSK can be generated. The most obvious way is to filter the modulating signal using a Gaussian filter and then apply this to a frequency modulator where the modulation index is set to 0.5, as shown in Figure 3.18. Whilst simple, this method has the drawback that the modulation index must exactly equal 0.5. In practice, this analogue method is not suitable because component tolerances drift and cannot be set exactly.

A second method is more widely used. Here, what is known as a quadrature modulator is used. The term 'quadrature' means that the phase of a signal is in quadrature, or 90°, to another one. The quadrature modulator uses one signal that is said to be in phase and another that is in quadrature to this. In view of the in-phase and quadrature elements, this type of modulator is often said to be an I–Q modulator (Figure 3.19). When using this type of modulator, the modulation index can be maintained at exactly 0.5 without the need for any settings or adjustments. This makes it much easier to use, and capable of providing the required level of performance without the need for adjustments. For demodulation, the technique can be used in reverse.

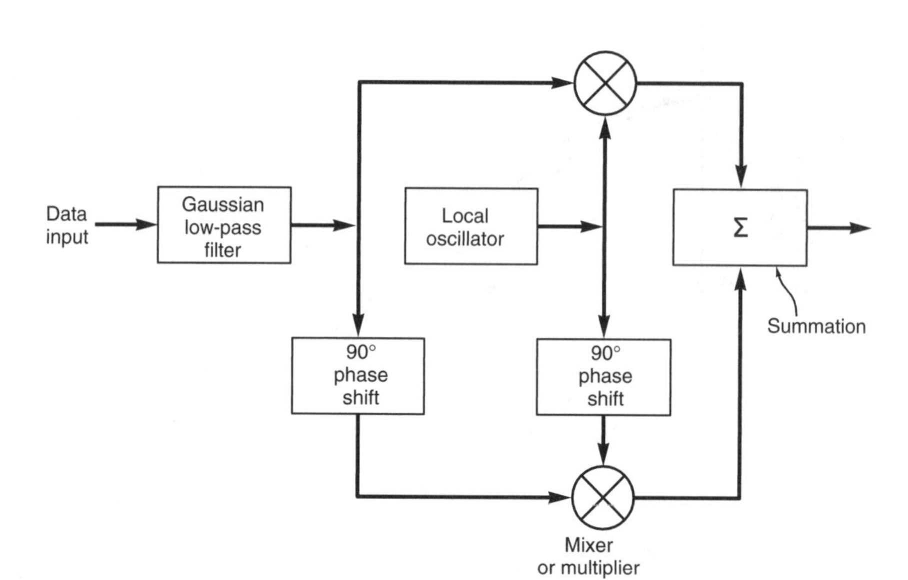

Figure 3.19 A block diagram of a quadrature or I–Q modulator used to generate GMSK.

A further advantage of GMSK is that it can be amplified by a non-linear amplifier and remain undistorted. This is because there are no elements of the signal that are carried as amplitude variations, and it is therefore more resilient to noise than some other forms of modulation.

Quadrature amplitude modulation

Another form of modulation that is widely used in data applications is known as Quadrature Amplitude Modulation (QAM). It is a signal in which two carriers shifted in phase by 90° are modulated, and the resultant output consists of both amplitude and phase variations. In view of the fact that both amplitude and phase variations are present, it may also be considered as a mixture of amplitude and phase modulation.

A continuous bit stream may be grouped into threes and represented as a sequence of eight permissible states:

Bit sequence	Amplitude	Phase (°)
000	1/2	0 (0°)
001	1	0 (0°)
010	1/2	$\pi/2$ (90°)
011	1	$\pi/2$ (90°)
100	1/2	π (180°)
101	1	π (180°)
110	1/2	$3\pi/2$ (270°)
111	1	$3\pi/2$ (270°)

Phase modulation can be considered as a special form of QAM where the amplitude remains constant and only the phase is changed. By doing this, the number of possible combinations is halved.

Although QAM appears to increase the efficiency of transmission by utilizing both amplitude and phase variations, it has a number of drawbacks. The first is that it is more susceptible to noise because the states are closer together, so that a lower level of noise is needed to move the signal to a different decision point. Receivers for use with phase or frequency modulation can both use limiting amplifiers that are able to remove any amplitude noise and thereby improve the noise reliance. This is not the case with QAM. The second limitation is also associated with the amplitude component of the signal. When a phase or frequency modulated signal is amplified in a transmitter there is no need to use linear amplifiers, whereas when using QAM that contains an amplitude component, linearity must be maintained. Unfortunately, linear amplifiers are less efficient and consume more power, and this makes them less attractive for mobile applications.

Spread spectrum techniques

In many instances it is necessary to keep transmissions as narrow as possible to conserve the frequency spectrum. However, under some circumstances it is advantageous to use what are known as 'spread spectrum techniques', where the transmission is spread over a wide bandwidth. There are two ways of achieving this: one is to use a technique known as frequency hopping, whilst the other involves spreading the spectrum over a wide band of frequencies so it appears as background noise. This can be done in different ways, and the two most widely used systems for this are DSSS and OFDM.

Frequency hopping

In some instances, particularly in military applications, it is necessary to prevent any people apart from intended listeners from picking up a signal or from jamming it. Frequency hopping may also be used to reduce levels of interference. If interference is present on one channel, the hopping signal will only remain there for a short time and the effects of the interference will be short lived. Frequency hopping is a well-established principle. In this system, the signal is changed many times a second in a pseudo-random sequence from a predefined block of channels. Hop rates vary, and are dependent upon the requirements. Typically the transmission may hop a hundred times a second, although at HF this will be much less.

The transmitter will remain on each frequency for a given amount of time before moving on to the next. There is a small dead time before the signal appears on the next channel, and during this time the transmitter output is muted. This is to enable the frequency synthesizer time to settle, and to prevent interference to other channels as the signal moves.

To receive the signal, the receiver must be able to follow the hop sequence of the transmitter. To achieve this, both transmitter and receiver must know the hop sequence, and the hopping of both transmitter and receiver must be synchronized.

Frequency hopping transmissions usually use a form of digital transmission. When speech is used, this has to be digitized before being sent. The data rate over the air has to be greater than the overall throughput to allow for the dead time whilst the set is changing frequency.

Direct sequence spread spectrum

Direct sequence spread spectrum (DSSS) is a form of spread spectrum modulation that is being used increasingly as it offers improvements over other systems, although this comes at the cost of greater complexity in the receiver and transmitter. It is used for some military applications, where it provides greater levels of security, and it has been chosen for many of the new cellular telecommunications systems, where it can provide an improvement in capacity. In this application it is known as Code Division Multiple Access, because it is a system whereby a number of different users can gain access to a receiver as a result of their different 'codes'. Other systems use different frequencies (Frequency Division Multiple Access – FDMA), or different times or time slots on a transmission (Time Division Multiple Access – TDMA).

Its operation is more complicated than those that have already been described. When selecting the required signal, there has to be a means by which the selection occurs. For signals such as AM and FM different frequencies are used, and the receiver can be set to a given frequency to select the required signal. Other systems use differences in time. For example, using pulse code modulation, pulses from different signals are interleaved in time, and by synchronizing the receiver and transmitter to look at the overall signal at a given time, the required signal can be selected. CDMA uses different codes to distinguish between one signal and another. To illustrate this, take the analogy of a room full of people speaking different languages. Although there is a large level of noise, it is possible to pick out the person speaking English, even when there may be people who are just as loudly (or maybe even louder) speaking a different language you may not be able to understand.

The system enables several sets of data to be placed onto a carrier and transmitted from one base station, as in the case of a cellular telecommunications base station. It also allows for individual units to send data to a receiver that can receive one of more of the signals in the presence of a large number of others. To accomplish this, the signal is spread over a given bandwidth. This is achieved by using a spreading code, which operates at a higher rate than the data. The code is sent repeatedly, each data bit being multiplied by each bit of the spreading code successively. The codes for this can be either random or orthogonal. Orthogonal codes are ones which, when multiplied together and then added up over a period of time, have a sum of zero. To illustrate this, take the example of two codes:

Code A	1	−1	−1	1
Code B	1	−1	1	−1
Product	1	1	−1	−1 summed over a period of time = 0, i.e. $1+1-1-1=0$

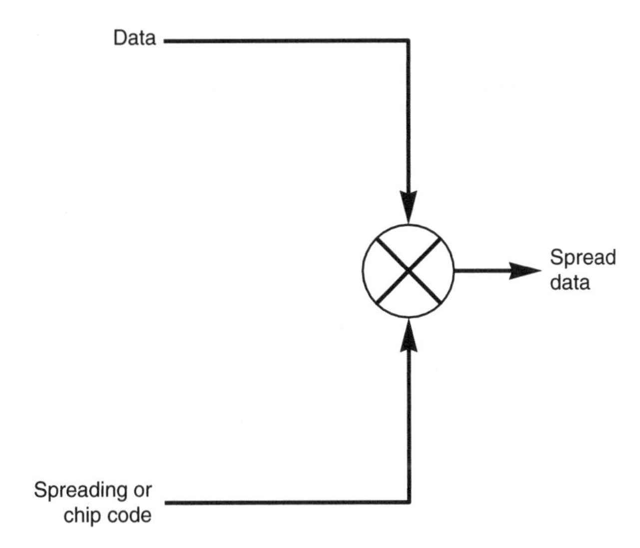

Figure 3.20 Multiplying the data stream with the chip stream.

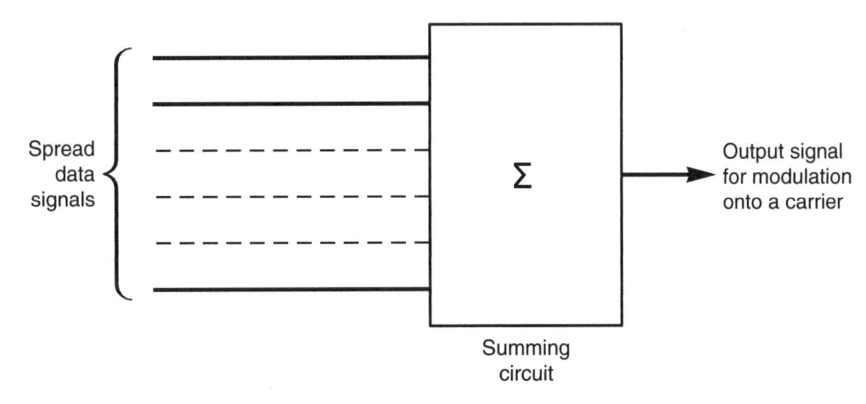

Figure 3.21 Generating a signal that carries several sets of data.

Using orthogonal codes, it is possible to transmit a large number of data channels on the same signal. To achieve this, the data are multiplied with the chip stream (Figure 3.20). This chip stream consists of the codes being sent repeatedly, so that the each data bit is multiplied with the complete code in the chip stream – in other words, if the chip stream code consists of four bits, then each data bit will be successively multiplied by four chip bits. It is also worth noting that the spread rate is the number of data bits in the chip code (i.e. the number of bits that each data bit is multiplied by). In this example the spread rate is four, because there are four bits in the chip code.

To produce the final signal that carries several data streams, the outputs from the individual multiplication processes are summed (Figure 3.21). This signal is then converted up to the transmission frequency and transmitted.

At the receiver, the reverse process is adopted. The signal is converted down to the base band frequency. Here, the signal is multiplied by the relevant chip code and the result summed over the data bit period to extract the relevant data in a process known as correlation. By multiplying by a different chip code, a different set of data will be extracted.

To see how the system operates, it is easier to refer to a diagram. In Figure 3.22, it can be seen that the waveforms (a) and (b) are the spreading codes. The spreading code streams are multiplied

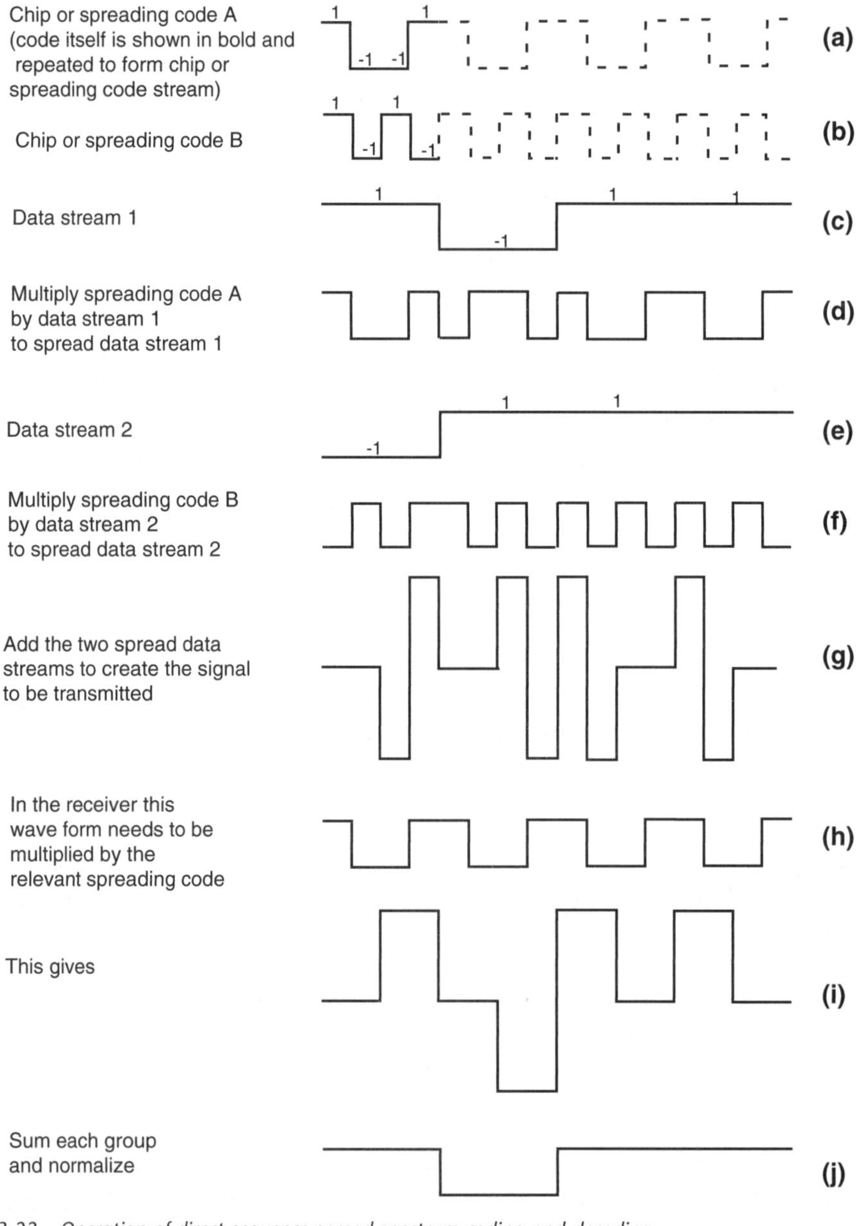

Figure 3.22 Operation of direct sequence spread spectrum coding and decoding.

with their relevant data. Here, the spreading code stream (a) is multiplied by the data in (c) to give the spread data stream shown in (d). Similarly, spreading code stream (b) is multiplied by the data in (e) to give (f). The two resulting spread data streams are then added together to give the baseband signal ready to be modulated onto the carrier and transmitted.

In this case, it can be seen that chip stream (a) is repeated in waveform (h). This is multiplied by (g) to give the waveform (i). Each group of four bits (as there are four bits in the chip code used in the example) is summed, and from this the data can be reconstituted as shown in waveform (j).

When a random or, more correctly, a pseudo-random spreading code is used, a similar process is followed. Instead of using the orthogonal codes, a pseudo-random spreading sequence is used. Both the transmitter and receiver will need to be able to generate the same pseudo-random code. This is easily achieved by ensuring that both transmitter and receiver use the same algorithms to generate these sequences. The drawback of using a pseudo-random code is that the codes are not orthogonal, and as a result some data errors are expected when regenerating the original data.

Orthogonal frequency division multiplex

Another form of modulation that is being used more frequently is Orthogonal Frequency Division Multiplex (OFDM). A form of this, known as coded OFDM or COFDM, is used for many Wi-Fi applications, such as IEEE Standard 802.11 as well as digital radio (DAB), and it is likely that it will be used for the fourth-generation (4G) mobile standards to provide very high data rates.

A COFDM signal consists of a number of closely-spaced modulated carriers. When modulation of any form – voice, data, etc. – is applied to a carrier, then sidebands spread out on either side. It is necessary for a receiver to be able to receive the whole signal in order to successfully demodulate the data. As a result, when signals are transmitted close to one another they must be spaced so that the receiver can separate them using a filter, and there must be a guard band between them. This is not the case with COFDM. Although the sidebands from each carrier overlap, they can still be received without the interference that might be expected because they are orthogonal to each another. This is achieved by having the carrier spacing equal to the reciprocal of the symbol period.

Figure 3.23 shows a traditional view of receiving signals carrying modulation, and Figure 3.24 shows the spectrum of a COFDM signal.

To see how this works, we must look at the receiver. It acts as a bank of demodulators, translating each carrier down to DC. The resulting signal is integrated over the symbol period to regenerate the data from that carrier. The same demodulator also demodulates the other carriers. As the carrier spacing equal to the reciprocal of the symbol period means that they will have a whole number of cycles in the symbol period, their contribution will sum to zero – in other words, there is no interference contribution.

One requirement of the transmitting and receiving systems is that they must be linear, as any non-linearity will cause interference between the carriers as a result of intermodulation distortion.

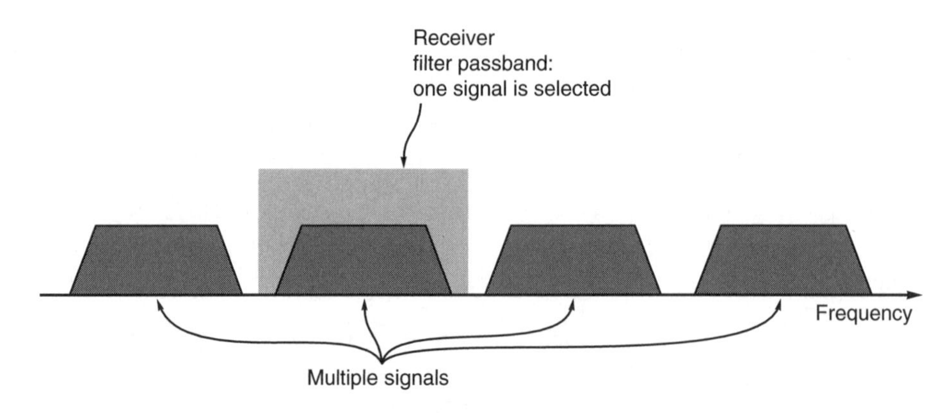

Figure 3.23 Traditional view of receiving signals carrying modulation.

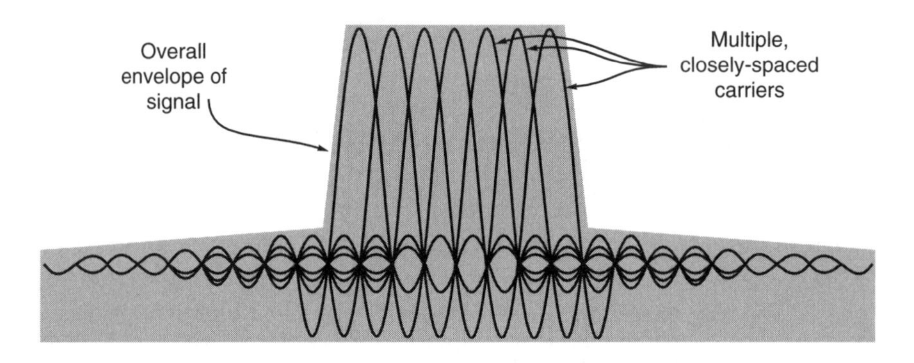

Figure 3.24 The spectrum of a COFDM signal.

This will introduce unwanted signals that will cause interference and impair the orthogonality of the transmission.

In terms of the equipment to be used, the high peak-to-average ratio of multicarrier systems such as COFDM requires the RF final amplifier on the output of the transmitter to be able to handle the peaks whilst the average power is much lower, and this leads to inefficiency. In some systems, the peaks are limited. Although this introduces distortion that results in a higher level of data errors, the system can rely on the error correction to remove them.

The data to be transmitted are spread across the carriers of the signal, each carrier taking part of the payload. This reduces the data rate taken by each carrier. The lower data rate has the advantage that interference from reflections is much less critical. This is achieved by adding a guard band time (or guard interval) into the system (Figure 3.25), which ensures that the data are only sampled when the signal is stable (i.e. a sine wave) and no new delayed signals arrive that will alter the timing and phase of the signal.

The distribution of the data across a large number of carriers has some further advantages. Nulls caused by multipath effects or interference on a given frequency only affect a small number of the carriers, the remaining ones being received correctly. Using error-coding techniques, which does

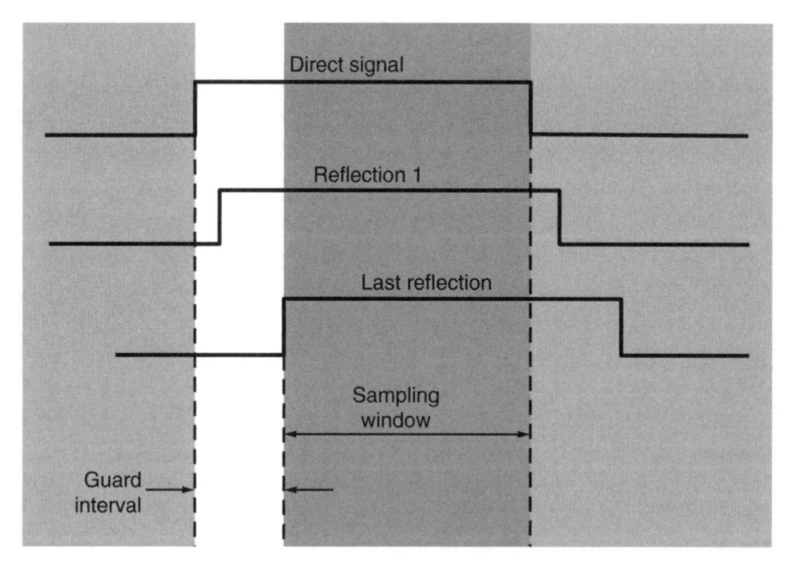

Figure 3.25 The guard interval used to prevent intersymbol interference.

mean adding further data to the transmitted signal, enables many or all of the corrupted data to be reconstructed within the receiver. This can be done because the error correction code is transmitted in a different part of the signal. It is this error coding that is referred to in the 'Coded' of COFDM.

Bandwidth and data capacity

One of the features that is of paramount importance in any communications system is the amount of data throughput. With users requiring more data and at faster rates, it is essential to make the optimum use of the available channels. It has been seen that there are many different types of modulation that can be used, some being more efficient than others for particular purposes. Nevertheless, there are certain laws that govern the amounts of data that can be transferred.

The bandwidth of a channel that is used is one of the major factors that influences the amount of data that can be accommodated. Bandwidth is literally the width of a band of frequencies measured in hertz (Hz). It is found simply by subtracting the lower limit of the frequencies used from the upper limit of the frequencies used.

Nyquist's theorem relates the bandwidth to the data rate by stating that a data signal with a transmission rate of $2W$ can be carried by frequency of bandwidth W. The converse is also true: given a bandwidth of W, the highest signal rate that can be accommodated is $2W$. The data signal need not be encoded in binary, but if it is then the data capacity in bits per second (bps) is twice the bandwidth in hertz. Multilevel signalling can increase this capacity by transmitting more bits per data signal unit.

The problem with multilevel signalling is that it must be possible to distinguish between the different signalling levels in the presence of outside interference – in particular, noise. A law known as Shannon's Law defines the way in which this occurs. It was formulated by Claude Shannon, a mathematician who helped build the foundations for the modern computer, and it is a statement of information theory that expresses the maximum possible data speed that can be obtained in a data channel. Shannon's Law says that the highest obtainable error-free data speed is dependent upon the bandwidth and the signal-to-noise ratio. It is usually expressed in the form:

$$C = W \log_2(1 + S/N)$$

where C is the channel capacity in bits per second, W is the bandwidth in hertz and S/N is the signal-to-noise ratio.

Theoretically, it should be possible to obtain between 2 and 12 bps/Hz, but generally this cannot be achieved and figures of between 1 and 4 bps/Hz are more reasonable. As a matter of simplicity, no attempt will be made here to provide a serious distinction between the two kinds of ways of measuring capacity and we will simply talk about 'bandwidth' in terms of bits per second. However, it must be remembered that bandwidth and digital data rate are two different quantities. Bandwidth is a measure of the range of frequencies used in an analogue signal, and bits per second is a measure of the digital data rate.

Error correction codes can improve the communications performance relative to un-coded transmissions, but no practical error correction coding system exists that can closely approach the theoretical performance limit given by Shannon's Law.

Summary

There are three ways in which a signal can be modulated: its amplitude, phase or frequency can be varied, although of these phase and frequency are essentially the same. However, there are a great many ways in which this can be achieved, and each type has its advantages and disadvantages. Accordingly, the choice of the correct type of modulation is critical when designing a new system.

Cellular basics

There are many different cellular radio or cellular telecommunications systems in use around the world. Naturally, they have many differences in the way they operate. Not only are there different implementations of similar technologies, but over the years there have also been developments in the technologies that are used. The first analogue systems gave way to digital systems, and in turn these are being migrated to technologies that are able to carry much higher data-rate signals to cope with the new applications that are being found. Yet despite these differences there are some fundamental concepts that are at the core of cellular telecommunications technology.

Spectrum re-use

One of the key requirements for any radio-based telecommunications system is the efficient use of the frequencies that are available. Early schemes for radio telephones used a single central transmitter to cover a wide area. These suffered from the limited number of channels that were available, with waiting lists for connection being many times greater than the number of people that were actually connected. This arose from the way the system was operated.

Take, for example, a system where each user is allocated a channel. Using a typical analogue system, each channel needs to have a bandwidth of around 25 kHz to enable sufficient audio quality to be carried, as well as allowing for a guard band between adjacent signals to ensure there are no undue levels of interference. Using this concept, it is possible to accommodate only forty users in a frequency band 1-MHz wide. Even if 100 MHz were allocated to the system, this would enable only 4000 users to have access to the system. Today cellular systems have millions of subscribers, and therefore a far more efficient method of using the available spectrum is needed.

The method that is employed is to enable the frequencies to be re-used. Any transmitter will have only a certain coverage area, and beyond this the signal level will fall to a limit below which it cannot be used and will not cause significant interference to users associated with a different transmitter (Figure 4.1). This means that it is possible to re-use a channel once outside the range of the transmitter. The same is also true in the reverse direction for the receiver, where it will be able

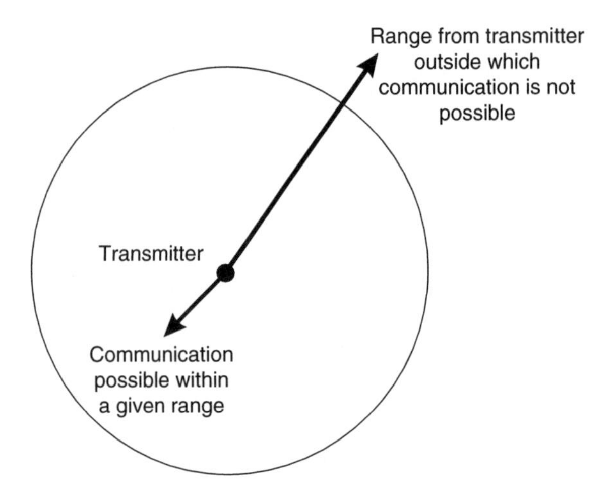

Figure 4.1 Range of a transmitter/receiver.

to receive signals over a given range only. In this way it is possible to split an area into several smaller regions, each covered by a different transmitter/receiver station.

These regions are conveniently known as cells. Diagrammatically, they are often shown as hexagonal shapes that conveniently fit together. However, in reality they have a very irregular outer boundary that is determined by the terrain and other factors. Additionally, it is not possible to define exactly the edge of a cell. The signal strength gradually reduces, and towards the edge of the cell performance falls. As the mobiles themselves also have different levels of sensitivity, this adds a further greying of the edge of the cell. It is therefore impossible to have a sharp cut-off between cells. In some areas they may overlap, whereas in others there will be a 'hole' in coverage.

To overcome this problem, in a basic cellular system adjacent cells are allocated different frequency bands so that they can overlap without causing interference. In this way, cells can be grouped together in what is termed a 'cluster'. Often these clusters contain seven cells, but other configurations are also possible (Figure 4.2). Seven is a convenient number for a variety of reasons. The first is that it gives sufficient isolation between cells where frequencies are re-used; this is necessary to ensure that the level of signals from nearby cells using the same frequencies is kept to an acceptable level. A further reason is that a group of channels is allocated to each cell. As there is a limit on the total number of channels available, it means that a sufficient number of channels can be allocated to each cell. The greater the number of cells in a cluster, the fewer channels can be allocated to each cell and the smaller its capacity. Conversely, when a smaller number of cells is included in each cluster the greater the number of channels that can be allocated to each cell, but then the separation from adjacent cells is reduced and levels of interference can rise.

Even though the number of cells in a cluster helps to govern the number of users that can be accommodated, by making all the cells smaller it is possible to increase the overall capacity of the network. However, a greater number of transmitter, receiver or base stations needs to be installed, thereby increasing the cost to the network operator. Accordingly, in areas where there are more users, small, low-power base stations are installed.

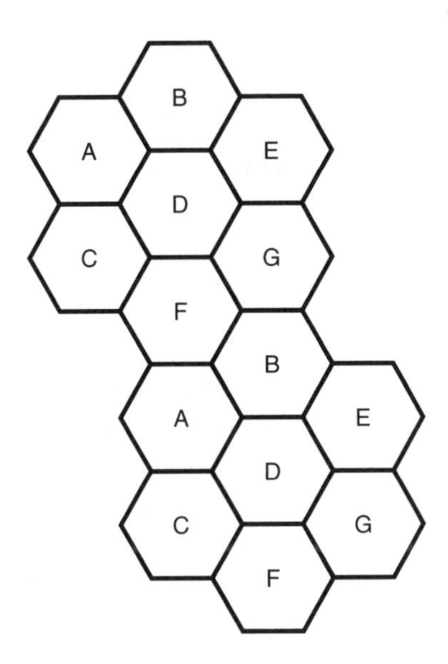

Figure 4.2 A typical seven-cell cluster. It is worth noting that although a hexagon is shown in this idealized diagram, the actual cell boundaries will be very irregular and will depend on the terrain and other factors.

The different types of cells are given different names. Macrocells are large cells that are generally used in remote or sparsely populated areas. These may be 10 km or possibly more in diameter. Microcells are those that are normally found in densely populated areas, and they may have a diameter of around 1 km. Picocells may also be used for covering very small areas, such as particular areas of buildings, or possibly tunnels where coverage from a larger cell is not possible. Obviously, for the small cells the power levels used by the base stations are much lower and the antennas are not positioned to cover wide areas. In this way, the coverage is minimized and the interference to adjacent cells is reduced.

Other types of cell may be used for some specialist applications. Sometimes 'selective cells' may be used where full 360° coverage is not required. They may be employed to fill in a hole in the coverage, or to address a problem such as the entrance to a tunnel. Another type of cell, known as an 'umbrella cell', is sometimes used in instances such as where a heavily used road crosses an area where there are microcells. Under normal circumstances this would result in a large number of handovers as people driving along the road quickly crossed the microcells. An umbrella cell can take in the coverage of the microcells while using different channels to those allocated to them. This would enable people moving along the road to be handled by the umbrella cell and thus experience fewer handovers than if they had to keep passing from one microcell to the next.

Multiple access schemes

Any cellular telecommunications system needs to have a method of allowing multiple subscribers to gain access to the system. These are known as multiple access schemes, and there are three main methods that are in use: Frequency Division Multiple Access (FDMA), Time Division Multiple Access (TDMA) and Code Division Multiple Access (CDMA).

The first of the schemes to be used was FDMA. This scheme is the most straightforward. As a subscriber comes onto the system, or swaps from one cell to the next, the network allocates a channel or frequency to each one (Figure 4.3). In this way, the different subscribers are allocated different slots and access to the network. As different frequencies are used, the system is naturally termed frequency division multiple access. This scheme was used by all analogue systems.

The second system came about with the transition to digital schemes. Here, digital data could be split up in time and sent as bursts when required (Figure 4.4). As speech was digitized, it could be sent in short data bursts; any small delay caused by sending the data in bursts would be short and go unnoticed. In this way it became possible to organize the system so that a given number of slots were available on a given transmission. Each subscriber would then be allocated a different time slot in which they could transmit or receive data. As different time slots are used for each subscriber to gain access to the system, it is known as time division multiple access. Obviously, this only allows a certain number of users access to the system. Beyond this another channel may be used, so systems that use TDMA may also have elements of FDMA operation as well.

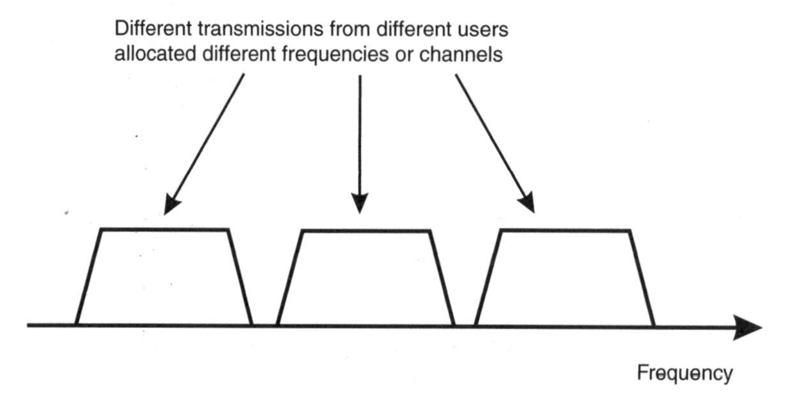

Figure 4.3 Diagram of an FDMA system.

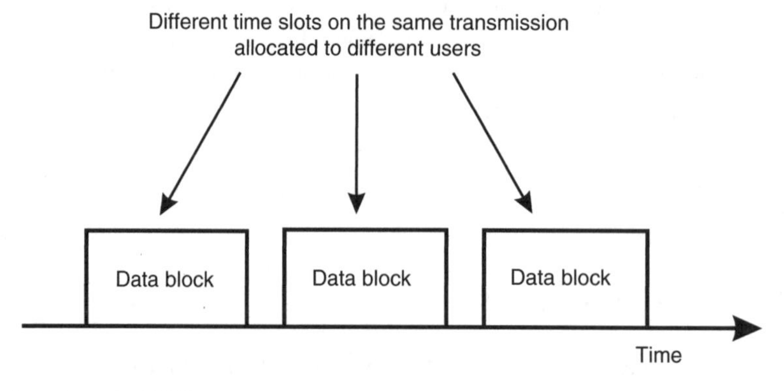

Figure 4.4 Operation of a TDMA system.

The final scheme employs one of the aspects associated with the use of direct sequence spread spectrum. In Chapter 3, it was seen that when extracting the required data from a DSSS signal it was necessary to have the correct spreading or chip code, and all other data from sources using different orthogonal chip codes would be rejected. It is therefore possible to allocate different codes to different users, and to use this as the means by which different users are given access to the system.

The scheme has been likened to being in a room filled with people all speaking different languages. Even though the noise level is very high, it is still possible to understand someone speaking in your own language. With CDMA, different spreading or chip codes are used. When generating a direct sequence spread spectrum, the data to be transmitted are multiplied with spreading or chip code. This widens the spectrum of the signal, but it can only be decoded in the receiver if it is again multiplied with the same spreading code. All signals that use different spreading codes are then not seen, and are discarded in the process. Thus, in the presence of a variety of signals it is possible to receive only the required one.

In this way the base station allocates different codes to different users, and when it receives the signal it will use one code to receive the signal from one mobile and another spreading code to receive the signal from a second mobile. In this way, the same frequency channel can be used to serve a number of different mobiles.

Duplex operation

Basic radio communications systems use a single channel and what is known as a 'press-to-talk' system, where the user presses a button (or 'pressel') on the microphone to talk and then releases the pressel to listen on the same frequency. This system is known as simplex, as it uses a single channel. For a phone system a full duplex system is required, where it is possible to speak in both directions at the same time. There are two main ways in which this can be achieved. The first is to transmit in one direction on one frequency, and simultaneously transmit in the other direction on another. To achieve this there must be sufficient frequency separation and filtering to ensure that the transmitter does not interfere with the receiver. A scheme that uses one frequency for transmitting traffic in one direction and another frequency for traffic in the other is known as Frequency Division Duplex (FDD; Figure 4.5a).

The other system uses only a single frequency, and can be employed where digital or data systems are used. This requires the analogue audio signal to be digitized. A single frequency is used for the radio frequency signal, and short packets of data are sent first in one direction and then the other. As these data bursts are relatively short, the user does not notice the brief delay introduced by the fact that the digitized speech signal is not sent immediately. This type of system is known as Time Division Duplex (TDD; Figure 4.5b).

It is often necessary to distinguish between the link from the mobile to the base station and that from the base station to the mobile. The former is often called the uplink or the reverse link, as the signal is being transmitted up to the base station. The latter is known as the downlink or the forward link.

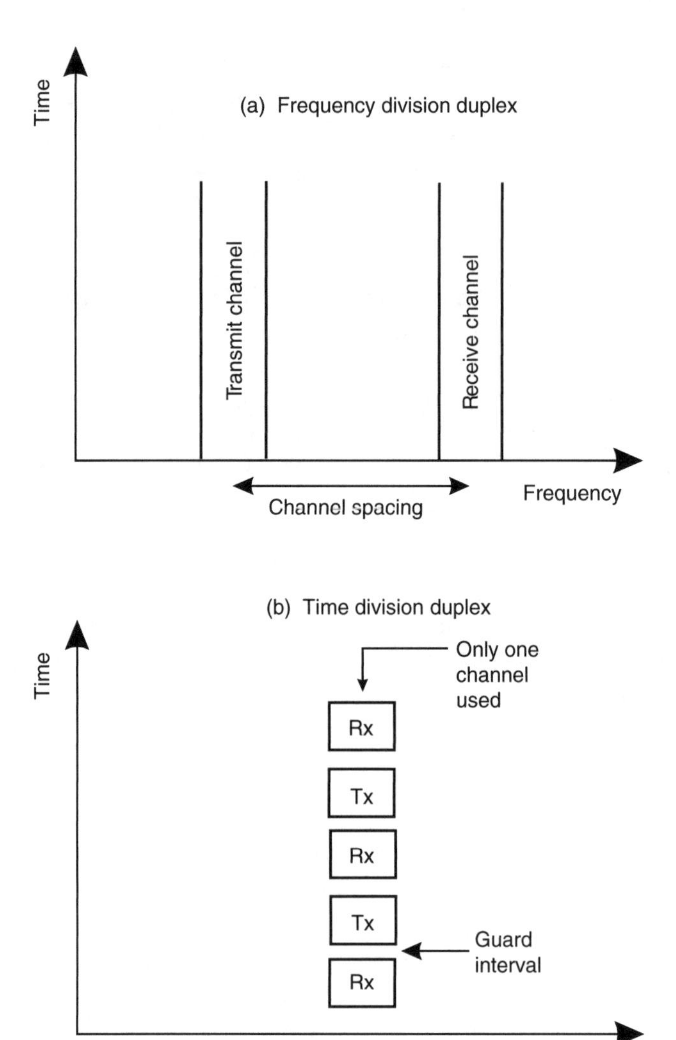

Figure 4.5 Representation of (a) frequency division duplex and (b) time division duplex.

Setting up calls

The process of setting up a call needs to be managed carefully. When the mobile is turned on it is not allocated a channel, time slot or chip code, dependent upon the type of access method used. Without being allocated a channel, it needs to be able to communicate with the base station. Also, calls need to be set up in both directions. In other words, it needs to be able to make a call if the user wants to call someone, and it also needs to be able to respond to incoming calls. To be able to respond to an incoming call there is a further complication, because the network needs to know where the mobile is in order to be able to route the call to the correct cell in order to contact the mobile in the first place.

There is a variety of tasks that need to be undertaken when a phone is turned on. It actually takes a few seconds from switching the phone on until it is ready for use. Part of this process is registration, and there are several aspects to this. The first is to make contact with the base station, and then the mobile has to register to allow it to have access to and use the network.

In order to make contact with the base station, the mobile uses a paging or control channel. The name of this channel and the exact way in which it works will vary from one cellular standard to the next, but it is a channel that the mobile can access to indicate its presence. The message sent is often called the 'attach' message. Once this has been achieved, it is necessary for the mobile to register.

There must be a register or database of users of a given network. With mobiles often being able to access all the channels available in a country, measures to ensure that the mobile registers with the correct network, and that the account is valid, are required. Additionally, this is required for billing purposes. To achieve this, an entity on the network, often known as the Authentication Centre (AuC), is used. The network and the mobile communicate, and numbers give the identity of the subscriber. The user information is then checked to provide authentication and encryption parameters that verify the user's identity and ensure the confidentiality of each call, protecting users and network operators from fraud.

Once accepted onto the network, two further registers are normally required: the Home Location Register (HLR) and the Visitors' Location Register (VLR). These two registers are required to keep track of the mobile so that the network knows where it is at any time in order for calls to be routed to the correct base station or general area of the network. These registers are used to store the last known location of the mobile. Thus at registration the register is updated, and then periodically the mobile updates its position. Even when the mobile is in what is termed its 'idle mode' it will communicate with the network on a regular basis to update its position and status. When the mobile is switched off, it sends a detach message. This informs the network that it is switching off, and enables the network to update its last known position.

Two registers are required – one for mobiles for which the network is the home network (i.e. the one with whom the contract exists) and the other for visitors. If there were only one register, then every time the mobile sent any message to the foreign network this would need to be relayed back to the home network, and this would require international signalling. The approach adopted is to send a message back to the HLR when the mobile first enters the new country, saying that the mobile is in a different network and that any calls for that mobile should be forwarded to the foreign visited network.

Receiving and making a call

One of the key elements in the receipt of an incoming call is the use of a paging channel. Under idle conditions, the mobile listens to the paging channel. When there is an incoming call, the network sends out a message in the cells in the area where the mobile is known to be located. The mobile monitors each message that is sent and compares the number with its own, responding only when the particular mobile in question has a call for which it is the destination.

Monitoring the paging channel all the time consumes current and this reduces battery life, although much less than when the mobile is transmitting. Battery life is of key importance to any mobile design, so many schemes to reduce battery drain and thereby prolong the battery life are implemented. As a result of this, many mobile systems adopt a procedure whereby the mobile has to listen for only part of the time.

One solution is to divide the paging messages into ten groups, depending on the last digit of the mobile telephone number. These groups, known as paging sub-groups, are sent in order and then repeated. This means that the mobile is required to synchronize with the timing of the paging message, and then it only needs to have the receiver active for 10 per cent of the time.

The drawback of the scheme is that it takes longer for calls to be routed through to the destination mobile. When a call needs to be set up, the network has to wait until it is time to transmit that particular paging group before it can send the paging message. This is the reason why it often takes some time to obtain a ringing tone after a mobile has been called. While this is a disadvantage, the gain of significantly improved battery life outweighs the drawback of the small additional wait to set up the call.

Receiving the paging message is only the first part of the exchange, as the mobile needs to acknowledge the call and the call needs to be set up. To achieve this, a channel that is often called the 'random access channel' is used. This channel serves two purposes, one is responding to paging messages, and the other is when the mobile needs to contact the base station to initiate a call. Mobiles send what is termed an 'access message', requesting a slot or channel for further communication.

The random access channel is given its name because mobiles access it randomly. As the network does not know when access is required, it cannot allocate slots. It is quite possible that two or more mobiles will access the channel at the same time, and both messages will collide. To ensure that the message has arrived and been read the mobile waits for an acknowledgement, and if no response is forthcoming then it waits a random amount of time before re-sending the message. By waiting a random length of time it significantly reduces the chances of a repeat collision with the same mobile. The chances of collisions are also reduced by keeping the messages on the random access channel short.

Once the message has been received an acknowledgement is sent, along with details of which channel to use. The mobile is then able to use the allocated channel to establish its communication.

Handover and handoff

The concept of a cellular phone system is that it has a large number of base stations each covering a small area of cells, and as a result frequencies are able to be re-used. Cell-phone systems also provide mobility. It is therefore a very basic requirement of the system that as the mobile handset moves out of one cell and into the next, it must be possible to hand the call over from the base station of the first cell to that of the second with no discernable disruption to the call. There are two terms for this process: 'handover' is used within Europe, and 'handoff' in North America.

Although the concept of handover is relatively straightforward, it is not an easy process to implement in reality. The system needs to decide when handover is necessary, and to which cell, and when the handover occurs it is necessary to re-route the call to the relevant base station while changing the communication between the mobile and the base station to a new channel. All of this needs to be undertaken without any noticeable interruption to the call. The process is quite complicated, and in early systems calls were often lost if the process did not work correctly.

Different cellular standards handle handover in slightly different ways, and a considerable amount of effort is placed on ensuring the standard defines a very robust method for handling handover, as this is such a key element in any cellular system. For the sake of explanation, the example of the way that GSM handles handover is given.

There are a number of parameters that need to be known to determine whether a handover is required – the signal strength of the base station with which communication is being made, and the signal strengths of the surrounding stations. The availability of channels also needs to be known. The mobile is obviously best suited to monitor the strength of the base stations, but only the network knows the status of channel availability, and the network makes the decision about when the handover is to take place and to which channel of which cell.

Accordingly, the mobile continually monitors the signal strengths of the base stations it can receive, including the one it is currently using, and it feeds this information back. When the strength of the signal from the cell the mobile is using starts to fall to a level where action needs to be taken, the network looks at the strength of the signals from other cells reported by the mobile. It then checks for channel availability, and if one is available it notifies the new cell to reserve a channel for the incoming mobile. When ready, the current base station passes the information for the new channel to the mobile, which then makes the change. Once there, the mobile sends a message on the new channel to inform the network that it has arrived. If this message is successfully sent and received, then the network shuts down communication with the mobile on the old channel, freeing it up for other users, and all communication takes place on the new channel.

Under some circumstances, such as when one cell is nearing its capacity, the network may decide to hand some mobiles over to another cell they are receiving that has more capacity, and in this way reduces the load on the cell that is running to near-capacity. Accordingly, access can be opened to the maximum number of users. In fact, channel usage and capacity are very important factors in the design of a cellular network.

Channel usage

One of the important factors in designing a cellular system is judging the usage that the system will receive. In order to see the greatest financial return, network operators are under pressure to optimize the usage of the system. However, when the system becomes over-used there are not sufficient channels to allow all the subscribers to gain access or, alternatively, when changing from one cell to the next there may be no free channels and the call is then dropped. Dropping calls is obviously not popular with subscribers, and if the levels of dropped calls rise too high then the subscriber is likely to move to another network, thereby increasing what is called the 'churn' rate (the rate at which subscribers move from one network to the next).

In order to judge the capacity requirement for a system it is necessary to look at the average time each subscriber is likely to use the system, the number of subscribers, and what the peaks in usage are likely to be. In general it is found that usage is high over most of the working day, rising slightly as people return home between about 5.00 and 6.00 pm.

In order to look at the traffic that is carried by a system, a unit called an Erlang (E) is used. The idea of analysing telephone usage was first investigated by Dr Erlang (hence the name), but the same ideas can be used for cellular telecommunications systems; the only difference is that radio is used as the medium to carry the signals, rather than copper wire.

In essence, 1 Erlang (E) is equivalent to one channel being used 100 per cent of the time. Thus if 100 channels are available and users are using them for 10 per cent of the time, then the capacity required is 10 E.

The amount of phone concentration (oversubscription) can be determined with the Erlang-B function (the Erlang blocked call function). Because the telephone network must be designed for the worst-case load, phone usage is defined as the level that is achieved during the busiest hour of the day.

There are ways of overcoming this. As the number of phone lines or channels increases, efficiency (in terms of fewer blocked calls) and oversubscription will also increase. This aligns with other similar examples, and the efficiency of all systems that use statistical multiplexing improves as the number of channel resources increases at the multiplexer. The concentration level moves from around 2:1 at ten subscriber phone lines to more than 3:1 at sixty user phone lines.

Infrastructure

The most obvious part of the cellular network is the base station. The antennas and the associated equipment often located in a container below are seen dotted around the country, and especially at the sides of highways and motorways. However, there is more to the network behind this, as the system needs to have elements of central control and it must also link in with the PSTN landline system to enable calls to be made to and from the wire-based phones, or between networks.

Different cellular standards, and in some cases different network operators, use different techniques. In fact, the early systems (and AMPS was one example) did not define the system beyond the air interface. Despite the differences between systems, the basic concepts are very similar. In any case, systems such as GSM have a well-defined structure, and this means that manufacturers' products can be standardized.

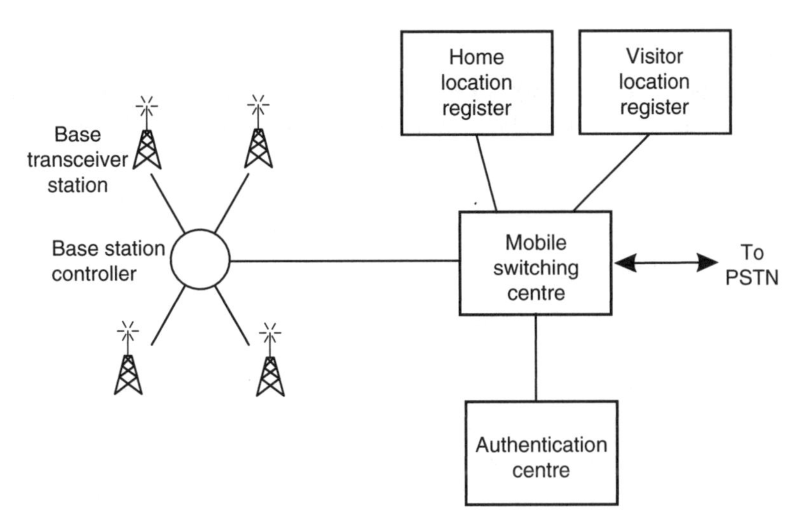

Figure 4.6 A typical network layout.

The overall system contains a number of different elements (see Figure 4.6), from the base transceiver station (BTS) itself with its antenna, back through a base station controller (BSC) and a mobile switching centre (MSC) to the location registers (HLR and VLR) and the link to the public switched telephone network (PSTN).

Of these units, the BTS provides the direct communication with the mobile phones. There may be a few base stations linked to a base station controller. This unit acts as a small centre to route calls to the required base station, and it also makes some decisions about which of the base stations is best suited to a particular call. The links between the BTS and the BSC may use either land lines or even microwave links. Often the BTS antenna towers also support a small microwave dish antenna used for the link to the BSC. The BSC is often co-located with a BTS.

The BSC interfaces with the mobile switching centre. This makes more widespread choices about the routing of calls and interfaces to the land-line based PSTN as well as the HLR and VLR.

Base transceiver station

The base transceiver station or system consists of a number of different elements. The first is the electronics section, normally located in a container at the base of the antenna tower. This contains the electronics for communicating with the mobile handsets, and includes radio-frequency amplifiers, radio transceivers, radio-frequency combiners, control, communication links to the BSC, and power supplies with back-up. Secondly, there is the antenna and the feeder to connect the antenna to the base station itself. These antennas are visible on the top of masts and tall buildings, enabling them to cover the required area. Finally, there is the interface between the base station and its controller further up the network. This consists of control logic and software, as well as the cable link to the controller.

The base stations are set up in a variety of places. In towns and cities the characteristic antennas are often seen on the top of buildings, whereas in the countryside separate masts are used (Figure 4.7). It is important that the location, height and orientation are all correct to ensure the required coverage is achieved. If the antenna is too low or in a poor location there will be insufficient coverage, leaving a coverage 'hole'. Conversely, if the antenna is too high and directed incorrectly, then the signal will be heard well beyond the boundaries of the cell. This may result in interference with another cell using the same frequencies.

The antenna systems used with base stations generally have two sets of receiver antennas. These provide what is often termed 'diversity reception', enabling the best signal to be chosen to minimize the effects of multipath propagation. The receiver antennas are connected to low-loss cable that routes the signals down to a multicoupler in the base station container. Here, a multicoupler splits the signals out to feed the various receivers required for all the RF channels. Similarly, the transmitted signal from the combiner is routed up to the transmitting antenna using low-loss cable to ensure the optimum transmitted signal.

Mobile switching centre

The MSC is the control centre for the cellular system, coordinating the actions of the BSCs, providing overall control, and acting as the switch and connection into the public telephone network. As such it has a variety of communication links into it, including fibre-optic links as well as some microwave links and some copper wire cables. These enable it to communicate with the BSCs, routing calls to them and controlling them as required. It also contains the Home and Visitor Location Registers – the databases detailing the last known locations of the mobiles – and the facilities for the Authentication Centre, which allows mobiles onto the network. In addition to this it also contains the facilities to generate the billing information for individual accounts.

In view of the importance of the MSC, it contains many back-up and duplicate circuits to ensure that it does not fail. Obviously back-up power systems are an essential element, to guard against the possibility of a major power failure, because if the MSC became inoperative then the whole network would collapse.

Mobile phone

The development of mobile phones has progressed tremendously since the first phones were launched on to the market in the 1980s. Not only has the size fallen dramatically, but the lifetime between charges has also risen, as has the number of facilities offered.

Although mobile phones are often sold at very low prices, this does not reflect the real price of their manufacture. They are highly sophisticated items of electronics equipment containing several different sections.

There is obviously the radio receiver and transmitter for communicating with the base station. Although the majority of mobiles that are sold and used these days transmit only low powers, the transmitter is still a highly complicated element of the circuitry, having to conform to very tight

Figure 4.7 A typical cellular telecommunications base station with its antenna tower.

limits for the level of spurious transmissions emitted. The transmitter power is dependent upon the system in use and the type of handset, but generally powers of only a few hundred milliwatts are transmitted. Even this may vary, as systems usually limit the power to what is actually required. In this way, interference levels are reduced and battery power is conserved. However, tight limits are sometimes placed on the power control tolerances, as in the case of GSM.

The choice of power amplifier mode of operation depends upon the system in use. Some systems do not require the use of a linear power amplifier, whereas others (like those using CDMA) need a relatively linear power amplifier. As linear power amplifiers are inherently less efficient than non-linear ones, there can be an issue with battery life.

The receiver is also of great importance. It must be sufficiently sensitive to enable it to operate under low signal conditions, but it must also be immune to the effects of other signals nearby – for example, when two mobile phones are operating close together. If the receiver were sensitive to strong off-channel signals, then its sensitivity could be significantly degraded when operating close to another mobile. This would clearly be unacceptable. Again, with the receiver battery drain is important. Not only must it have a low level of power consumption, but most receivers also have an idle state when they periodically monitor channels to check for paging information. By turning off the majority of the receiver, the drain on the battery can be reduced.

The antenna is another significant item. Today many antennas are contained within the body of the phone, although a number of phones do have small antenna positions. In either case, the size of the antenna has to be limited and this places significant constraints on the design. In addition to this, the antenna must be capable of operating in the proximity of nearby objects. Not only is the phone itself nearby, but the hand of the person using the phone may also be shrouding it. In all cases, the antenna design must be able to operate adequately under these conditions.

There is a significant level of control circuitry within a phone. This undertakes a large number of tasks. It obviously controls the transmitter and receiver, setting them to the right channels as instructed by the network. It also controls the changeover from transmit to receive. In addition to this it will put various sections of the circuitry into a sleep mode when they are not required, to limit the current drain on the battery and conserve power.

Any communication with the network has to be in the correct format. Data sent to and from the base station need to be correctly formatted. There is also a large number of protocol messages that are sent to ensure that the network and mobile operate correctly together. Messages such as attach, detach, those sent when initiating and completing a call, when handing over to another base station, and many more all need to be correctly controlled, and this is done by the control circuitry.

The man–machine interface (MMI), consisting of the keypad and display, also needs to be managed so that the user can operate the phone. This too is undertaken by the control circuitry. However, the MMI also provides many other functions, including the phone book, text message organization, games, clock, and many more of the applications that are finding their way into modern mobile phones.

Another part of many phones these days is what is called the SIM card on GSM phones. It is called the USIM (Universal Subscriber Identity Module) for UMTS and the RUIM (Removable User Identity Module) on CDMA phones.

The battery is a major element of the phone. Its power capacity governs the time for which the phone can operate between charges. In recent years, battery technology has improved considerably. Initially nickel cadmium power sources were used, which can be detrimental to the environment if disposed of incorrectly. The next source was the nickel metal hydrogen battery, which provided virtually the same cell voltage as the NiCad battery but with a slight improvement in performance. Most of all, these cells did not contain the harmful cadmium constituent. The next development of cell technology came with the introduction of the lithium ion (or Li-ion) battery. This offers a high voltage of around 3 volts per cell, and also a greatly increased power density. This enables the cells to be made much smaller while providing a similar or possibly larger capacity. In the future, it is likely that fuel cells may be used. Rather than having to recharge them, these cells require re-fuelling. With the varieties likely to be used in cell phones running on methanol or ethanol, the refills will be cheap and easy to use.

Voice coding

All the second- and third-generation mobile phone systems use digital techniques in various forms to transmit the voice signals of a phone call. This requires that the analogue voice signals have to be converted to a digital format.

The conversion into a digital format is accomplished using an analogue-to-digital converter (ADC). The ADC samples the waveform at fixed time intervals, and generates a binary representation of the value of the waveform at that instant. The actual voice quality is dependent upon two basic factors, namely the sampling rate and the number of bits in each sample.

It is possible to determine the minimum sampling rate from Nyquist's theorem. This states that the sampling rate must be twice the maximum frequency of the waveform. If a straight conversion were used, then the amount of data required to be sent would be too high to be accommodated in a cellular phone channel. If the maximum frequency of the audio were 4 kHz, this would require a sampling rate of 8 kHz, and if there were 8 bits per sample, then it would require a data rate of 64 kbps.

To overcome this problem, vocoders are used. These are able to analyse and compact the data representing the speech so that it requires a considerably lower data rate. Typically, data rates below 15 kHz are attainable. To achieve this level of performance they use complex algorithms to analyse the speech, and the speech is then reconstituted at the receiving end.

A variety of different vocoders is used, many of which use a technique known as Code Excited Linear Prediction (CELP). The choice is dependent upon the cellular standard in use and, in some instances, the conditions. However, they all give significant improvements in the data rate and bandwidth required to transmit the signals.

Digital data structures

With cellular systems focusing on digital technology these days, the data that are transmitted have to be carefully organized. In this way the receiver, whether in a mobile handset or a base station, knows what the data being received relate to, and can thus handle them accordingly.

Typically, data are organized into three elements for TDMA systems. These are frames, slots and channels, and they follow this hierarchy. The channels should not be confused with the radio-frequency channels that are used; they are in effect data channels that are contained within the overall data stream.

A frame is what may be termed an all-inclusive data package. It is made up of a number of slots which are in a given order, and the overall structure is repeated after a given time. Thus the frame carries control information indicating features about the frame, its length etc; it also carries the payload (typically this will be data contained within the slots), and error-checking information. There may be start bits as well for synchronization.

The slots contain individual information destined for or arriving from a particular mobile, and are a result of the fact that different users are allocated different time slots. A given user is allocated a particular time slot, and in this way the relevant data can be inserted into or extracted from the right time slot.

The next level down the hierarchy consists of the channels. These handle call-processing elements, and are the sections of data required for particular purposes. For example, channels may be used for actions including paging, call initiation, control, traffic data and the like.

Within CDMA systems that are not time-multiplexed, slots are not used; instead, the different channels are made up by using different spreading codes on the same radio-frequency carrier. In this way, the different functions, types of data and users can be separated from one another.

Analogue systems

The first mobile phone systems to be launched were based on analogue technology. By today's standards the performance was limited, the phones were large, and initially the coverage was poor. In addition to this, when the phone systems were first launched they were very expensive, which limited them to business use. Nevertheless, these phone systems marked a major milestone in telecommunications history. For their day they represented a major step forward in technology, and they also established a market that grew well beyond the initial predictions. In fact, many of these systems quickly ran out of capacity as a result of the demand as the numbers of subscribers grew.

Around the world a number of different standards were introduced. Although many of the systems were adapted from others, it still meant that incompatible systems were often in use from one country to the next. As roaming was not a concept that had gained a foothold by this time this was not a major problem, but it did mean that manufacturers could not gain the full benefits of scale that might otherwise have been possible, even though many of the differences were relatively small.

The three main systems that were launched were NMT (Nordic Mobile Telephone), AMPS (Advanced Mobile Phone System, also known under its specification number as IS-41) and TACS (Total Access Communications System). The NMT system was developed as a joint venture by Nordic countries with Ericsson, and Nokia playing a major part, although there were many other interested parties. This system was obviously used in Nordic countries, but was also employed elsewhere around the world. AMPS was focused on North America, but was also widely used around the globe. Similarly TACS, developed by Motorola, was used in the UK as well as many other countries.

All these systems used FM for the voice channel. The channel spacing for AMPS was 30 kHz, for TACS it was 25 kHz and for NMT it was 25 kHz. A further system, known as NAMPS or Narrowband AMPS, was also introduced with the aim of conserving spectrum, but this did not gain widespread acceptance because the first digital systems were in the process of being developed by this time. In general the systems used frequencies in the region of 900 MHz, although the NMT system made use of an allocation around 450 MHz.

The major analogue mobile phone systems are summarized in Table 5.1.

Table 5.1 Summary of the major analogue mobile phone systems.

	UK	North America	Scandinavia	Japan	West Germany
System Transmission frequency (MHz)	TACS	AMPS	NMT	NTT	C450
Base station	935–960	869–894	463–467.5	870–885	461.3–465.74
Mobile	890–915	824–849	453–457.5	925–940	451.3–455.74
Transmit/receive spacing (MHz)	45	45	10	55	10
Channel spacing (kHz)	25	30	25	25	20
Number of channels	1000	832	180	600	222
Audio modulation	FM	FM	FM	FM	FM
Frequency deviation (± kHz)	9.5	12	5	5	4
Control signal modulation	FSK	FSK	FSK	FSK	FSK
Control signal frequency deviation (+/−kHz)	6.4	8	3.5	4.5	2.5

An extension to the TACS system was introduced to provide further spectrum and thereby provide further capacity. The system was known as Extended TACS, or ETACS. The original phones did not provide access to these channels, so the network allocated channels in the new band first to those phones with ETACS capability. In this way maximum flexibility was maintained. All other standards of the original specification were maintained; the ETACS extension purely provided additional spectrum to accommodate the rapidly growing demand.

Today

Most of the analogue systems have been withdrawn in favour of the second generation digital systems. However, analogue systems are still in use in many outlying areas of North America as they provide good service for these areas. Accordingly, the focus on this chapter will be on AMPS rather than TACS or any of the other analogue systems. However, the basic concepts for all the systems are very similar, the main differences being in the parameters for the air interface.

Basic system

The basic concept of an analogue cellular phone system is very similar in many ways to that of the other second- and third-generation systems, save that it is much simpler. There is also a high degree of similarity between the different types of analogue systems. Essentially they are complete telephone networks. The base stations communicate directly with the mobiles, and a group of mobiles is placed under the control of a base station controller. This will provide local control for the network and undertake many of the actions that can be performed locally.

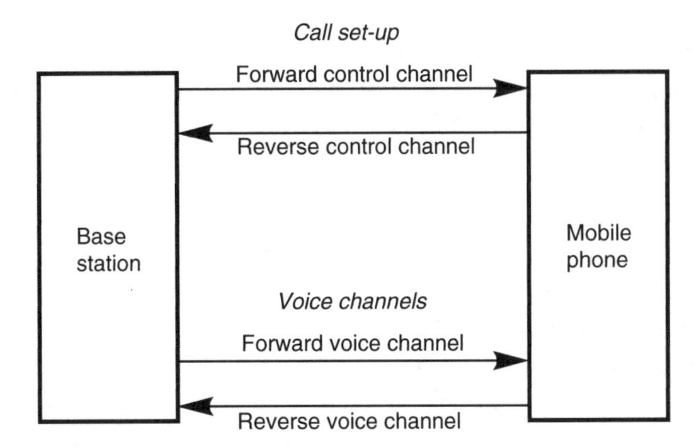

Figure 5.1 The channels used in communicating with mobile stations.

Table 5.2 AMPS channel frequency bands.

	Reverse channel bands (MHz)	Forward channel bands (MHz)
A (initial)	824–834	869–879
A (Extended)	844–846.5	889–891.5
B (Initial)	834–844	879–889
B (Extended)	846.5–849	891.5–894

The system utilizes two types of physical channel to communicate with the mobiles: the voice channel and the control channel. Both are present in the forward and reverse directions. As seen from Table 5.1, the voice channels use frequency modulation to carry the voice traffic, whereas the control channel uses frequency shift keying as this channel transmits data. There is one pair of traffic channels (i.e. forward and reverse) per user, and one pair of control channels per base station (Figure 5.1). The control channels are primarily used to set up and manage the calls, sending information such as the channels to be used when a call is initiated and those to be used when a handover is effected. For NMT, some channels have reversible roles.

For the AMPS system the channels are split into two, namely A and B, and then allocated to different operators. The original allocations were extended to provide more capacity after the popularity of the cellular phone systems became evident. These are not contiguous, as seen in Table 5.2, and accordingly the channels are numbered 1 to 1024.

Base station

Typically base stations are allocated between twenty and thirty voice channels. In addition to this they have a single signalling or control channel that carries all the paging and access functions for the cell. Typically the base stations are organized in the classic seven- or twelve-cell arrangement to provide adequate channel re-use, while still maintaining sufficient levels of separation to ensure that interference levels are reduced to satisfactory figures. In some urban areas where system usage is very high the cell coverage areas are reduced, and the cells may be arranged in a four-cell repeat pattern with six sectors to a cell. Each of the sectors is then allocated its own set of voice channels, although a single signalling channel is retained for the whole base station.

The voice channels and signalling channel outputs are combined using a high Q cavity resonator. As the system operates in full duplex and the transmission mode is FM for the voice channels and FSK for the paging or control channel, this is on all the time. The output from the combiner is fed to the antennas. Often these are of the collinear variety, providing around 9 dB gain over a dipole. The radiation is omni-directional to provide coverage all around the cell. In some instances, the directional pattern may be altered to ensure that interference between cells using the same channels is reduced to the minimum.

For receiving, many base stations are fitted with six directional antennas. These have considerably more gain than the transmitting antenna, and they are often mounted at 60° angles. The outputs from these antennas are then switched so that the optimum signal can be chosen for the receiver. To ensure that the receiver is always provided with the best signal, a scanning receiver is used. This monitors every channel on the base station on every antenna every few seconds, and ensures that the best signal is routed through to the relevant receiver for each channel. In this way, the optimum reception is maintained.

The use of multiple antennas with additional gain for receiving while maintaining an omni-directional transmit pattern has several advantages. The base station is able to transmit a higher power level than the mobile, and therefore antenna gain is less important. However, for receiving, the fact that the mobile is able to transmit only a lower power can be compensated for at the base station by having a high-gain antenna. The gain in the antenna means that it will be directional, and this is overcome by having a set of antennas with the scanning receiver.

The network and base station are provided with identification codes. In the USA, the System Identifier (SID) is assigned by the Federal Communications Commission (FCC) along with the carrier frequencies for each geographical area. The system identifier is used by the mobile to ensure that it is on the correct network.

Mobile switching centres

In a network there may typically be around twenty mobile switching centres. These mobile switching centres (MSC) or, as they are sometimes called in AMPS, Mobile Telephone Switching Offices (MTSO), are effectively a digital exchange, and they have a distributed architecture. To achieve this, the switching centres are linked together to provide a fully integrated network. Also, the base station controllers are linked to the most convenient MSC, often by a dedicated line, although other links are used as appropriate. The base station controllers typically control a small number of individual base stations and are linked either by a wire landline or, often, a short-range microwave link.

Mobile equipment

By today's standards, the mobile phones used for analogue systems were large and offered a lower level of performance. Battery life was much shorter. This was a result of the fact that semiconductor technology was less advanced, and frequency modulation was used. This requires that the full power be transmitted all the time that a call is in process (one of the advantages of the TDMA technology used for the second-generation GSM and NA-TDMA technologies is that the duty cycle during a call is only 12.5 per cent). Furthermore, both transmit and receive functions are required during a call and this means that the two frequency synthesizers used as the local oscillators are required – one for the transmit side of the phone and the other for the receive side. This again increased battery consumption.

Mobiles are given a Mobile Identifier Number (MIN). This is equivalent to a land-line telephone number, and includes an area code identifying the home service area, a three-digit exchange number and a four-digit subscriber identification number. This number is used in a variety of areas, not least of which is to route calls through to the mobile phone.

The mobile also has what is termed an electronic serial number (ESN). This code is permanently assigned at manufacture and acts as a security number, helping to prevent fraud – particularly in the form of phone cloning. The code is 32 bits long, and consists of three fields. These include an 8-bit manufacturer code, an 18-bit serial number and 6 bits reserved for future use. Many manufacturers have tried to implement a feature that renders the phone inoperative if there is any attempt to alter the ESN.

A further identifier within the phone is able to inform the network of the capabilities of the phone. Parameters such as the number of channels it can support, the power output class and the like can be passed to the network.

Voice messaging

The voice is carried on the main 'traffic' channel, one of which is allocated to each mobile making a call. The audio is modulated onto the carrier using FM, which causes the carrier to deviate in line with the instantaneous level of the modulation. The actual level of deviation varies between the different systems, but for AMPS the maximum deviation is ±12 kHz, whereas for TACS it was 9.5 kHz.

The audio is processed to ensure that the carrier signal is used as effectively as possible. It is found that speech consists of short high peaks, with longer periods of low intensity. This is fine when listening to someone talking face to face, but for an electronic medium such as a mobile phone it leads to considerable inefficiency. If not processed, the system would have to be set to enable it to cater for the high peak levels, but this would mean that the average level of deviation was very low and thus the speech would become very susceptible to any noise present. It was found that if the average level of the speech could be raised, it would be much easier to hear and less susceptible to noise.

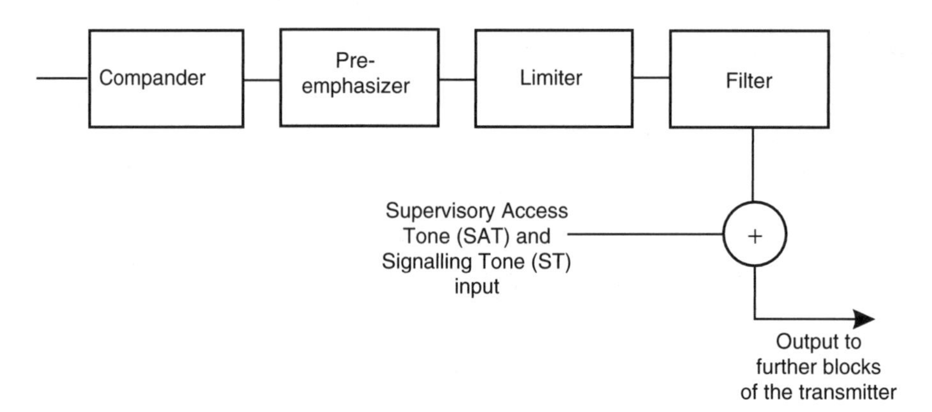

Figure 5.2 Audio chain in a mobile.

A number of processes are used to achieve the required level of voice intensity (Figure 5.2). The first is what is termed a 'compander'. This is a compressor/expander. At the transmitter end of the link the dynamic range of the voice is compressed, and at the receiver it is expanded. In this way a much higher level of average modulation can be applied to the signal, and by expanding the signal at the receiver it is not unduly distorted. For AMPS, a 2:1 syllabic compander is used. This operates on a syllable-by-syllable basis, and for every 2-dB change in input signal the output only varies by 1 dB.

In addition to the compander, another process known as pre-emphasis is used in the transmitter. Similarly, a de-emphasis filter is used in the receiver. This process raises the level of some of the higher-frequency sounds during transmission, and reduces them at the receiver. The reason for this is to reduce the background hiss that would otherwise be discernible.

Two further processes are applied. The deviation of the signal needs to be kept within its defined limits to ensure that the signal does not deviate into the adjacent channel and cause interference. Careful attention is paid to this. Being a frequency-modulated signal, both the frequency of the modulating signal and its amplitude are important. The first stage in the process is to pass the signal into a limiter. This removes any peaks in the signal that would cause the deviation limit to be exceeded. Although it may sound as though this might cause severe distortion to the signal audible to the call recipient, this is not the case. Finally, the signal is passed through a low-pass filter. This removes frequencies above about 3 kHz, ensuring that the audio signal does not exceed its maximum bandwidth and thereby cause the bandwidth of the overall signal to become too wide. The filter is placed after the limiter to ensure that harmonics introduced by the limiter that fall outside the required bandwidth are removed.

Signalling methods

In general, the control channels are used to page or call a mobile and also take access requests from a mobile to connect to a network or initiate a call. The actual format of the data that are transmitted varies between one system and the next. As a result of the fact that AMPS is still in use in some areas, this system is the focus of the following explanation. The same basic principles can be applied to other systems.

Data are transmitted using frequency shift keying at a rate of 10 kbps with Manchester coding.

It is also possible to send some signalling over the voice channels. This is achieved by using a Supervisor Audio Tone (SAT), which is generated by the base station. This is transmitted with a small level of deviation, typically around 1.7 kHz, but with a frequency of around 6 kHz. The fact that the frequency is above the audio range means that it can be filtered out of the audio stream and not heard by the caller, and the low level of deviation means that the overall bandwidth of the radio signal remains within its limits. The SAT frequency must be transmitted to a tolerance of 1 Hz. It is filtered by the receiver in either the base station or a mobile, and when detected it can be used to decode the relevant information.

One of the main purposes of the SAT is to ensure that the mobile is communicating satisfactorily with the correct base station. As it is possible that distant base stations may be received from time to time, the SAT can be used to ensure that communication is established with the correct one. As a result, clusters of base stations are allocated a Supervisory audio tone Colour Code (SCC). Where a channel frequency is re-used, a different colour code is employed.

During call set-up the base station allocates a mobile voice channel, and it also informs it of the correct SCC to expect. The base station commences sending the SAT on the forward voice channel, so that the mobile can detect it as it receives the channel. If it does, the SAT signal will pass through the audio filter correctly, indicating to the mobile that it is receiving the signal from the correct base station. It will then respond by placing a SAT on its voice channel which will be expected by the base station. In this way the SAT enables a form of 'handshaking' to take place without using up time on the control channel.

The SAT signal is also used after a call has been set up. It is particularly important during handovers, and operates in much the same fashion as described above. In addition to this it is used as a basic acknowledgement signal, for example when confirming commands from the base station and to ensure the integrity of the link between the mobile and base station.

If a SAT is not detected, or the incorrect one is detected, than a Fade Timing Status (FTS) flag is set. This enables a 5-second count after which the call base station transmission is discontinued, assuming a lost call.

Table 5.3 Supervisor audio tone frequencies.

SAT frequency (Hz)	SCC
5070	00
6000	01
6030	10

A further tone, known as the Signalling Tone (ST), is used in AMPS. Unlike the SAT, which is generated by the base station and then relayed back by the handset, the ST is actually generated within the handset and has a frequency of 10 kHz. Thus tone is used in combination with the SAT to provide an indicator of a variety of states during the course of a call. It is also used as an 'acknowledge' signal in some instances.

Control channels

The control channels are the means by which the radio elements of the cellular system are organized. As such the control channels are of prime importance, maintaining order within the radio channels and informing the mobiles which channels they need to use.

For AMPS, there is a total of twenty-one channels that are reserved in each direction. One control channel is assigned to each base station, and mobiles can search for the strongest control channel to assess the best base station to access.

There are two types of control channel: forward and reverse. For AMPS, the forward channel is known as the FOrward Control Channel (FOCC) and the other as the REverse Control Channel (RECC).

Forward control channel

The FOCC is transmitted continuously by the base station so that it can be received by all mobiles. It also acts as a beacon for mobiles joining the network, and for others looking for alternative base stations in case they need to handover from one base station to the next. With a total of twenty-one control channels available, mobiles scan these frequencies to look for the strongest signal, as the one with the strongest control channel is likely to provide the best overall communications.

It is interesting to note that not only can the antenna patterns be altered to change the coverage area of the base station, but also adjustment of the power of the control channel has a similar effect. If the power is reduced then the range of the signal is reduced, and mobiles in a more restricted area will find it as the strongest signal. Similarly, the receive sensitivity and thresholds need to be altered so that the transmitter and receiver are compatible.

The control channel obviously performs other functions as well. It transmits control data to the mobiles so that they can act in accordance with the wishes of the network. Data is transmitted at 10 kbps, and it consists of frames of 463 bits which take 46.3 ms to transmit.

The data stream is made up from several components. It starts with synchronization or sync bits, after which there are two words. Each of these words is repeated five times in a frame. The words are addressed to different mobiles. Mobiles with even MINs refer to word A, and those with odd MINS refer to word B. In addition to this there are busy/idle bits. These are sent every 10 bits, and are used to control the access of the mobile to the system.

Reverse control channel

The RECC uses a 48-bit word which is repeated five times, making a 240-bit sequence. The mobiles transmit on the channel only when instructed to do so, or when accessing the system. As a result, there may be times when it appears to be unoccupied.

There are three main messages that are carried by the RECC, and these are registration, origination and page response. The registration message is sent by the mobile before any calls are set up. The mobile needs to be accepted onto the network prior to either receiving or sending any calls. Origination messages are also required. These are sent when a number is dialled and the mobile communicates this with the network. Finally, responses are required when a mobile is paged to say that there is an incoming call.

Call initiation

In order to make a call from a mobile phone, the phone has to originate the call and communicate its intention to make a call. The phone knows the correct channel to use from information broadcast on the FOCC, and transmits a request on the channel.

As the system does not have access to the network, it cannot ask for a slot in which to transmit so it will not interfere with other mobiles. Accordingly, it uses a technique known as Carrier Sense Multiple Access (CSMA). Using this scheme, the mobile phone sends its message and waits for an acknowledgement. If it does not receive an acknowledgement, it assumes that another mobile is transmitting and has blocked its message. It waits until the other mobile should have stopped transmitting and tries to send its message again, repeating this until it receives a response.

The message from the mobile contains information about it, including its MIN and ESN. After transmitting the message, it listens on the FOCC for the acknowledgement and channel assignment. The base station forwards the request to the Mobile Switching Centre and, after validating the mobile, selects a traffic channel pair for the mobile. If no channels are available, the network rejects the request.

If the network allows the mobile onto the network, it allocates the channels and sends this information to the mobile, which then retunes its transmitter and receiver accordingly, and the SAT is used to confirm that all is set correctly. Additionally the network routes the call to the PSTN, where it is further routed through to the destination, or to another mobile on the network.

All call and power control from this point forward is handled using SAT and ST on the AMPS traffic channel assigned to the mobile.

Paging and incoming call set-up

When an AMPS mobile is not engaged in a call, it monitors the FOCC for any paging calls that may occur. The phone checks the correct data words in the data frames sent out on the paging channel.

As explained previously, mobiles with even MINs listen to one of the data words, and those with odd MINs listen to the other.

When a call attempt is directed to the mobile, the paging signal is set more than once. It is actually repeated after several seconds in case the mobile was temporarily in an RF 'hole' or otherwise unable to receive the first page. The time interval between pages is short to minimize the ringing delay experienced by the originator of the call.

Once received, the mobile responds to the page using the RECC. It then awaits the network assigning it a pair of traffic channels. The mobile response is also repeated, in case the initial response collided with another mobile on the reverse control channel or the message does not succeed as a result of a poor signal path or interference.

Once the mobile has received details of the channels that are assigned to it, it proceeds to these channels. At the same time, the network produces the audible ringing tone and the phone starts to ring. As in the case of a mobile-originated call, all further signalling takes place using the tones on the voice or traffic channel.

Handoff

During the course of a call it is quite likely that the mobile will change its location, and the signal strengths between the mobile and the base station and the base station and the mobile will fall as the mobile moves away from the base station. During the call the scanning receiver monitors the quality of the received signal in terms of what is essentially a signal-to-noise ratio. If this measurement falls below a predetermined level, the base station sends a message to the mobile switching centre to advise of the situation.

The system them initiates a search for a more suitable cell by requesting the adjacent cells to measure the strength of the signal of the relevant mobile.

If the signal is received better in another base station, a new channel on that station is allocated if one is available. A second voice path is then set up and bridged across to the existing one in preparation for the handoff.

Once this is complete, the system or network generates a handoff instruction and sends it over the FOCC. This carries information that includes the new channel to be used along with the SAT. The mobile accepts the information and sends an ST signalling tone for 50 ms. It then turns off the reverse channel transmission and retunes to the new channel. The transmitter in the mobile is turned off so that the signal does not sweep across all the other channels in between, causing interference.

Once it reaches the new channel, the mobile turns its voice transmitter channel transmission on and retransmits the SAT of the new base station. This acts as confirmation that the new link has been established. Accordingly, when the new base station receives this it informs the mobile switching centre that the handoff has been successfully accomplished. It then informs the old base station to release the channel the mobile was previously using, making it available for further use.

Summary

The analogue systems were a major technological achievement for their time, and in addition to this they also established a market that grew very much faster than expected. However, there were several drawbacks to the system that were soon discovered. With the rapidly growing number of subscribers, capacity started to become a major issue. This was why the FCC in the USA, and the equivalent in governments around the world, started to allocate further spectra. In the UK, the TACS system was extended as ETACS to provide further capacity. However, it was obvious that new technology would be needed to overcome the problem in the long term.

Another problem was that of security. Not only was it possible to listen in to conversations, but in addition to this it was possible to clone phones very easily.

As a result of these problems, new systems were developed to ensure that capacity and quality of service targets could be met, and security levels significantly improved. The resulting systems used digital technology and resulted in the migration from the first-generation analogue systems to the second-generation digital ones.

GSM

The system now referred to as GSM started life as the Groupe Speciale Mobile. Later this was changed to represent its global nature, while keeping the same initials, to Global System for Mobile communications.

The aim was an initiative to develop a digital mobile phone system that would support roaming across country borders and also provide improved capacity when compared to the analogue systems that were then in use. As the press had made much of the lack of security of some of the analogue systems that had been in use, with more than one high-ranking dignitary having their calls listened to by outsiders, call privacy and security were also high on the agenda.

The goals set for the system were tough, as no digital system had been launched before and neither had any attempts at roaming on this scale been considered. The original aim was to use spectra in the 900-MHz region, but since its initial launch further bands have been used at 1800, 1900 and, more recently, 800 MHz. Originally the systems using the high-frequency bands were given the names DCS 1800 (Digital Communications System) and PCS 1900 (Personal Communications System). Now they are simply referred to as GSM 1800 and GSM 1900 respectively.

The system has been a considerable success, far outstripping the take-up by any other network beforehand or currently available. In early 2004 the system had over a billion subscribers, and this figure has continued to rise.

Part of the success of the GSM system is that the standard has rigorously defined the whole system. The standard was generated and controlled under the auspices of ETSI. In this way a truly international standard was generated, and all manufacturers and service providers have to adhere to the standard. In this way, not only did global roaming become a reality; it also meant that manufacturers could cater for a global market and benefit from economies of scale. These benefits of cost could also be passed on to the consumer, making it a very attractive solution. Previously there had been a large variety of different systems, and manufacturers had to tailor their solutions to the relevant countries and systems. Even now the other systems in use around the world are not standardized to the same degree, and this prevents seamless roaming across countries and even different networks within the same country. The running of the GSM standard has now been passed to 3GPP, the organization that is also controlling the third generation UMTS (Universal

Mobile Telecommunications System) that uses wideband CDMA. In this way it is possible for the GSM and UMTS developments to be more closely linked.

System architecture

The architecture of the GSM system has been proven to be very successful, many of the names and ideas being adopted by other systems. Its basic concepts are also incorporated in the new UMTS/W-CDMA 3G system as well, although many of the names are slightly different to prevent confusion.

The main elements of the system are the Base Transceiver Station (BTS), the Base Station Controller (BSC), the Mobile Switching Centre (MSC) and the registration and authentication areas (Figure 6.1). These include the Home Location Register (HLR), the Visitor Location Register (VLR), the Equipment Identity Register (EIR) and the Authentication Centre (AuC).

The BTS is the primary communication area with the mobiles. The BTS transmits and receives the signals from them handling the interface protocols. The BTS is then linked to a BSC, which controls a small group of BTSs. These two are linked using an interface known as the A-bis interface. Like all the other interfaces between elements of the GSM network, this is

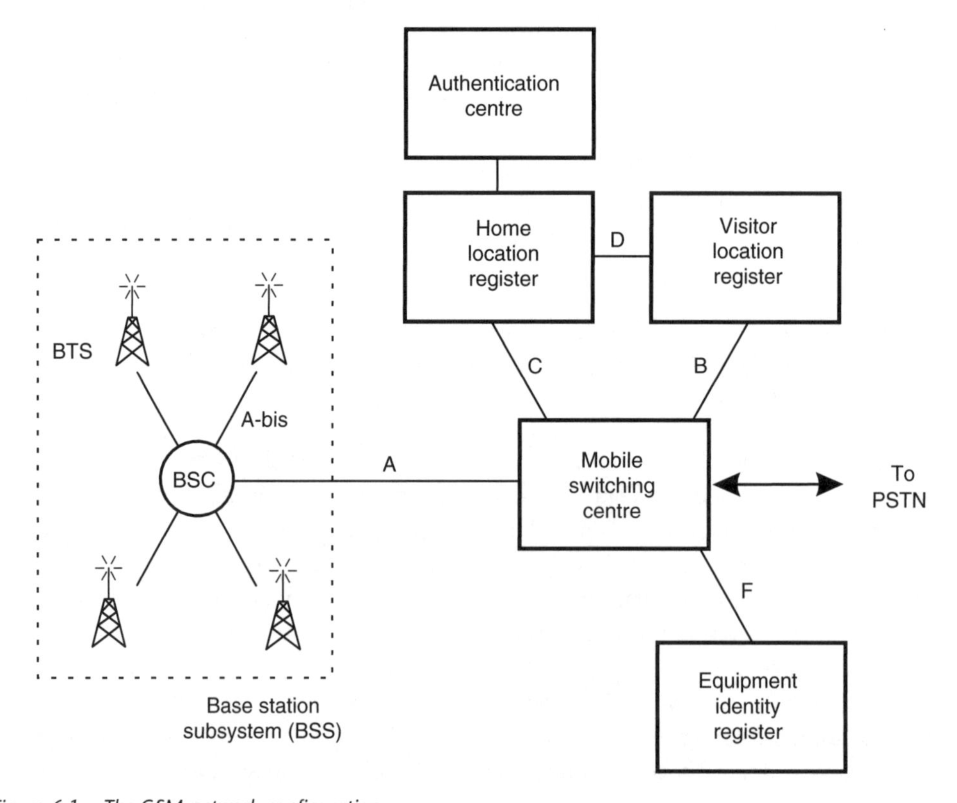

Figure 6.1 The GSM network configuration.

now rigorously defined to enable equipment from different manufacturers to be used together. However, some of this standardization came late, and it means that many older components are proprietary makes, enabling only BSCs and BTSs from particular manufacturers to be connected.

The BSC manages one or more BTSs. It handles the radio channel set-up, frequency hopping control and handovers. It links with the MSC or mobile switching centre via an interface termed the A interface. A further task normally handled by the BSC is to translate the 13-kbps voice data rate used over the radio link to the standard 64 kbps used by the PSTN.

The core of the network sub-system is the MSC, or mobile switching centre, acting like a switching node of the PSTN. In addition to this it interfaces with the AuC to provide authentication allowing users onto the network. It also interfaces with the HLR and VLR to provide location information for the network, so that calls can be routed to the correct BTS – including those that may need to be routed overseas to mobiles that are roaming. In addition to this, it coordinates handovers.

Equipment and subscriber identifiers

There are several different identifiers that are incorporated within the GSM standard to provide flexibility while still being able to retain the required level of security. A number of these identifiers are stored on a card, known as the Subscriber Identity Module (SIM) card. This small memory card is inserted into a mobile to provide information about the subscriber. SIM cards also contain other information, including numbers in the 'phone book', thereby enabling subscribers to change the mobile equipment while still retaining their call number and other information. Other identifiers are stored within the mobile equipment itself, or on the network.

International mobile equipment identity

The IMEI is a fifteen-digit number used to identify the equipment. Entered into the mobile at manufacture, it is checked when a mobile tries to access a network and compared with the Equipment Identity Register (EIR) database of the numbers of known stolen mobiles. The status returned in response to an IMEI query to the EIR may fall into one of three categories: white-listed, in which case it is allowed to connect; grey-listed, in which case it remains under observation by the network for possible problems; and black-listed, when it is either reported stolen or is not a type-approved mobile, and is therefore not allowed to connect.

International mobile subscriber identity

The IMSI is a fifteen-digit number that is contained within the SIM card and allocated to the user by the network operator. It enables the operator to link the phone number and the subscriber. The number includes a country identifier code to enable the item to be operated globally.

Temporary mobile subscriber identity

The TMSI is assigned by the network and is shorter than the full IMSI address, thereby making it more efficient to send.

Authentication key and cipher key

The authentication key (Ki) is stored on the SIM card and is used to compute the cipher key (Kc), which is used in the encryption algorithm to prevent unauthorized listening to the mobile message.

Air interface

The GSM system uses digital TDMA technology combined with a channel bandwidth of 200 kHz. In this way it is able to offer a higher level of spectrum efficiency than that achieved with the previous generation of analogue systems. As there are many carrier frequencies available (124 at 900 MHz for the basic frequency allocation, although there are extensions to provide additional channels), one or more can be allocated to each base station.

The system uses frequency division duplex, and as a result the channels are paired – one for the downlink from the BTS to the mobile and another for the reverse link back to the BTS. The frequency difference between the two channels varies according to the band in use. For 900 MHz there is a difference of 45 MHz between transmit and receive, when the 1800-MHz band is used the frequency difference is 95 MHz, and for the 1900 MHz band the difference is 80 MHz (Table 6.1). In addition to the frequency difference, transmit and receive functions take place at different times, using different time slots, and this significantly eases the problem of interference between the transmitter and receiver. It also significantly simplifies the electronics, as some areas of the RF chain can be used for both transmit and receive functions. One example of this is the frequency synthesizer used as the local oscillator, which can change between transmit and receive functions and frequencies, thereby requiring only one section of hardware rather than repeating the circuitry for both the transmit and receive paths.

The carrier is modulated using Gaussian Minimum Shift Keying (GMSK), and the data rate for the overall carrier is approximately 270 kbps. GMSK is used for the GSM system because it is able to provide features required for GSM. It is resilient to noise when compared to some other forms of modulation, it occupies a relatively narrow bandwidth, and it has a constant power level.

The relative immunity to interference and noise allows for increased re-use of the available spectrum between cells, thereby allowing a higher channel capacity for a given channel quality.

Table 6.1 Transmit and receive bands for the various GSM bands.

Band/system	BTS transmit (mobile receive)	BTS receive (mobile transmit)
900 MHz	935–960 MHz	890–915 MHz
DCS 1800	1805–1880 MHz	1710–1785 MHz
PCS 1900	1930–1990 MHz	1850–1910 MHz

The narrow bandwidth and spreading of the carrier enables the optimum use to be made of the spectrum, by ensuring that adjacent channels do not need a large guard band between them to make sure that interference levels are kept to acceptable limits. Finally, the fact that the output signal level is constant enables more efficient power amplifiers to be used. This enables the power amplifiers in the mobiles to consume less power for a given output level, thereby lengthening battery life.

The data transported by the carrier serve up to eight different users. Even though the full data rate on the carrier is approximately 270 kbps, some of this supports the management overhead, and therefore the data rate allotted to each time slot is only 24.8 kbps. In addition, error correction is required to overcome the problems of interference, fading and the like. This means that the available data rate for transporting the digitally encoded speech is 13 kbps for the basic vocoders.

Each cell site normally has several different radio carrier frequencies. One of these frequencies for each cell site has a single time slot allocated as a control channel, and this is known as the beacon frequency. This is time shared, but if activity (especially that of SMS) increases then further time slots may be allocated.

Power levels

A variety of power levels is allowed by the GSM standard. The highest is 20 watts (43 dBm) and the lowest is 800 mW (29 dBm). As mobiles may transmit for only one-eighth of the time (i.e. for their allocated slot, which is one of eight), the average power is one-eighth of the maximum.

Additionally, to reduce the levels of transmitted power and hence the levels of interference, mobiles are able to step the power down in increments of 2 dB from the maximum to a minimum of 13 dBm (20 mW). The mobile station measures the signal strength or signal quality (based on the bit error rate), and passes the information to the BTS and hence to the BSC, which ultimately decides if and when the power level should be changed.

A further power-saving and interference-reducing facility is the discontinuous transmission (DTx) capability that is incorporated within the specification. It is particularly useful because there are long pauses in speech – such as when the person using the mobile is listening – and during these periods there is no need to transmit a signal. In fact, it is found that a person speaks for less than 40 per cent of the time during a normal telephone conversation. The most important element of DTx is the Voice Activity Detector, which must correctly distinguish between voice and noise inputs – a task that is not trivial. If a voice signal is misinterpreted as noise, the transmitter is turned off and an effect known as clipping results. This is particularly annoying to the person listening to the speech. However, if noise is misinterpreted as a voice signal too often, the efficiency of DTx is dramatically decreased.

It is also necessary for the system to add background or comfort noise when the transmitter is turned off, because complete silence can be very disconcerting for the listener. Accordingly, this is added as appropriate. The noise is controlled by the Silence Indication Descriptor (SID).

Multiple access and channel structure

GSM uses a combination of both TDMA and FDMA techniques. The FDMA element involves the division by frequency of the (maximum) 25-MHz bandwidth into 124 carrier frequencies spaced 200 kHz apart as already described.

The carriers are then divided in time, using a TDMA scheme. The fundamental unit of time is called a 'burst period', and it lasts for approximately 0.577 ms (15/26 ms). Eight of these burst periods are grouped into what is known as a TDMA frame. This lasts for approximately 4.615 ms (i.e.120/26 ms), and forms the basic unit for the definition of logical channels. One physical channel is one burst period allocated in each TDMA frame.

In simplified terms, the base station transmits two types of channel; traffic and control. Accordingly the channel structure is organized into two different types of frame, one for the traffic on the main traffic carrier frequency and the other for the control on the beacon frequency. The frames are grouped together to form multiframes, and in this way it is possible to establish a time schedule for their operation so the network can be synchronized.

The traffic channel frames are organized into multiframes consisting of twenty-six bursts and taking 120 ms, and control channel multiframes that comprise fifty-one bursts and occupy 235.4 ms. In a traffic multiframe, twenty-four bursts are used for traffic. These are numbered 0 to 11 and 13 to 24. One of the remaining bursts is then used to accommodate the SACCH described later, the remaining frame remaining free. The actual position used alternates between positions 12 and 25.

Multiframes are then constructed into superframes taking 6.12 seconds. These consist of fifty-one traffic multiframes or twenty-six control multiframes. Above this 2048 superframes are grouped to form one hyperframe, which repeats every 3 hours, 28 minutes and 53.76 seconds (see Figure 6.2).

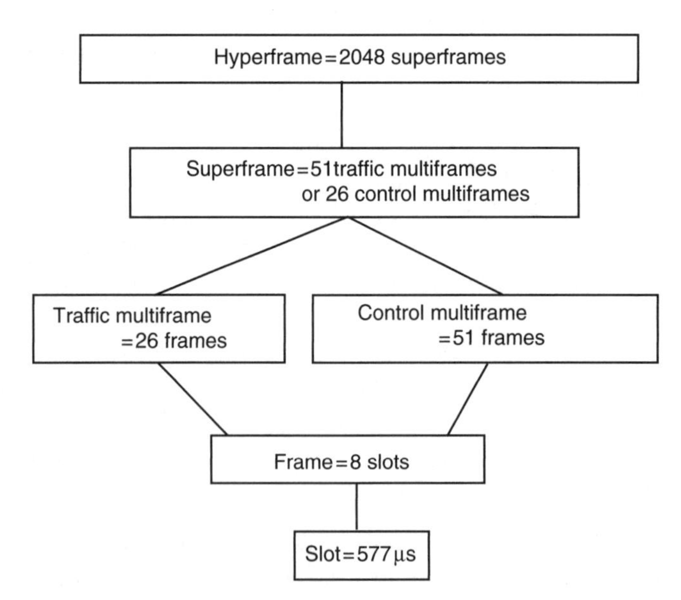

Figure 6.2 GSM frame organization structure.

Control channels

There is a variety of different control channels that are used to provide the required functionality to enable the mobiles and the BTS to communicate, set up and manage the calls. There are a number of channels, and they can be split into groups, namely broadcast channels for initial synchronization, common control channels for initiating calls, and dedicated control channels to manage calls. The common channels include the following.

1. *Broadcast channels*:

   - Frequency Correction Channel (FCCH). This is effectively a series of 148 zeros sent as the first logical channel in the control sequence. As the burst is all zeros, this equates in GMSK to a sine wave at a frequency of approximately 67 kHz. The mobile finds this and adjusts its own frequency to be that of the network.
   - Synchronization Channel (SCH). This is transmitted in time slot zero in the frame following FCCH. It contains extended training sequences to allow the mobile to gain exact synchronization with the network. Also contained within the sequence are the BTS identification and the current frame location in relation to the hyperframe.
   - Broadcast Control Channel (BCCH). This channel is continually broadcast on the downlink, i.e. by the BTS. It contains information including base station identity, frequency allocations, and frequency-hopping sequences that are needed by the mobiles to set up and receive calls. One broadcast control channel segment occupies four frames per multiframe.

2. *Common control channels*:

   - Paging Channel (PCH). This channel is used to 'page' or call the mobile to let it know that there is an incoming call. The location registers are able to identify the relevant cell for the mobile, and a request is sent out for a particular mobile (on the paging channel that all mobiles monitor periodically) to inform the mobile that there is an incoming call.
   - Access Grant Channel (AGCH). This channel is used to set up a call once a mobile has been paged on the paging channel or random access channel. It directs the mobile to another control channel, typically the Slow Dedicated Control Channel (SDCCH), where the call set up is progressed.
   - Random Access Channel (RACH). This performs an equivalent function to that of the paging channel, except that it is used when a mobile initiates a call. Accordingly, it is located within the uplink from the mobile to the BTS. As the name implies, the channel is accessed randomly as required by the mobile. As any mobile communicating with the BTS can access this channel, there is the possibility that two mobiles will interfere with one another. To overcome this problem, the BTS sends an acknowledgement to the mobile once it has correctly accepted the access. If this does not occur, the mobile waits a random amount of time before re-accessing via the RACH. On acceptance of the access the BTS directs the mobile to the SDCCH, a two-way channel where the call can be set up.

3. *Dedicated control channels*:

- Standalone Dedicated Control Channel (SDCCH). This channel is used for messaging along with the companion SACCH in order to relay signalling information. The channel takes four bursts in each multiframe for each mobile that is active on the BTS, which is just sufficient to enable timely transmission of the relevant data.
- Slow Associated Control Channel (SACCH). This channel is used to relay a variety of information. On the downlink from the BTS to the mobile, this includes broadcast messages which give the beacon frequencies of neighbouring cells, power control information, and timing advance. For the uplink it includes the measurement report, which gives the strength measurements from signals received from neighbouring cell beacon transmissions, acknowledgement for the power control and acknowledgement for the timing advance.
- Fast Associated Control Channel (FACCH). This channel is used to send urgent messages that are normally unscheduled. As the SACCH may take seconds to transfer, this may not be swift enough for many functions. In order to achieve much faster data transfer, the messages are sent in place of speech data for short periods of time. Additionally, it comes free when a TCH is activated, so part way through a call set-up signalling will move to FACCH, not because it needs to be fast or is unscheduled, but because it is an available and efficient resource. It is referred to as 'fast' because it can carry up to fifty messages per second against four per second for SDCCH.

It is often possible to take up to about one in six of the traffic or speech frames without undue loss of quality, although there is some degradation. When a frame of speech data is lost as a result of the presence of an FACCH frame being sent, the same action is taken as when data is lost due to poor signal or channel conditions. Here, the receiving end replaces the lost speech data with a repeat of the previous sound. In this way the listener hears a persistence of the sound rather than a silence or click, and in this way the missing data are masked sufficiently. As the number of frames that are used by the FACCH increases, the degradation of the speech deteriorates. This happens in a non-linear fashion as the number of frames increases.

It is necessary that the receiver knows whether a frame is traffic (i.e. speech or FACCH) so that the data can be used correctly. To achieve this, the data are marked in each burst using two flag bits. These are set to 0 for user or speech data and to 1 for FACCH data. Two data bits are required because it is possible that the FACCH and speech data may be interleaved where the data is spread between adjacent bursts, and only half the data may be speech and half FACCH.

The traffic channel

As the name implies, the Traffic Channel (TCH) is used to carry the voice information for the call. There are three basic types of traffic channel: full-rate, half-rate and eighth-rate. A full-rate traffic channel (TCH/F) occupies a complete slot per frame, whereas a half-rate traffic channel (TCH/H) occupies only every other burst of a time slot. The eighth-rate traffic is used only on the SDCCH for the exchange of call set-up or the short message service (SMS).

There are eight different time slots that are allocated, a mobile on the network normally transmitting during one of them. Similarly, the BTS transmits to a particular mobile during one burst. In some special instances, such as sending high data rates, it is possible for a mobile to be allocated more than one burst period. Also, when transmitting speech using a half-rate vocoder the mobile may use only one in sixteen burst periods.

The burst periods are allocated numbers 0 to 7. The system is also organized so that equivalent bursts in the uplink and downlink do not occur at the same time (see Figure 6.3). In this way the mobile does not transmit and receive simultaneously, thereby simplifying the radio transmitter receiver circuitry considerably. Therefore, when the BTS is transmitting slot 0 it is receiving slot 5 – i.e. the mobile allocated to slot 5 is transmitting.

The channel is made up of a variety of different types of data, which are grouped together logically into 'bit fields'. These bit fields include the initial tail bits, training sequences, flag bit final bits and the data payload, and there is also a guard time between bursts to or from different mobiles.

The first 3 bits within any burst are set aside to enable the transmitter power to increase to the required level. This is required because it is not possible for the transmitter to turn on instantly. Also, if it were to come on prematurely to be ready for its burst, then it would interfere with the end of the transmission from the previous mobile.

Following the initial tail bits, a flag is sent followed by a section of data consisting of 57 bits, and it is normally digitized voice. On some occasions this may be replaced by signalling, where a channel called the FACCH is incorporated. The type of data, whether digitized voice or FACCH, is indicated by the flag.

In the centre of the burst there is a group of data 26 bits long, called the training sequence or midamble. This is used as a timing reference and as an adaptive equalizer training pattern. There are eight patterns (chosen to be as orthogonal as possible) used within GSM, each 26 bits long, and the same one is used in all eight time slots. Different patterns are used in cells to ensure

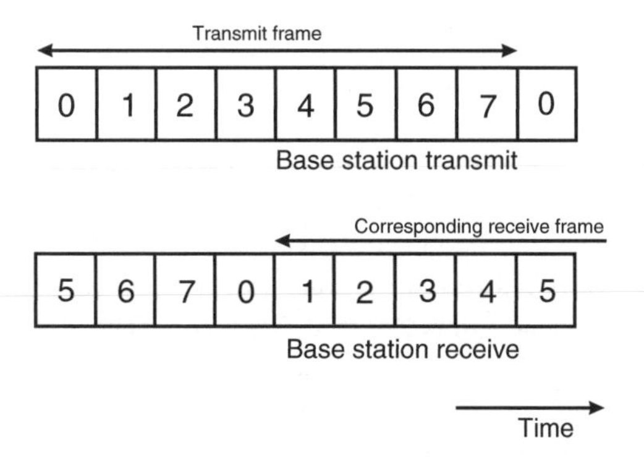

Figure 6.3 Time offsets on corresponding frames for uplink and downlink.

that only the signal from the required BTS is recognized. With the movement of mobiles, signals vary considerably, and it is possible that the signal from a nearby cell may be received suddenly. The use of the training patterns enables the mobile to detect this.

The data, usually from the voice encoding, need to be split to fit the data structure. The speech is broken into blocks of 20 ms, and this is processed by the speech coder to give 260 bits to represent the speech. These are ordered and divided into two groups. For full-rate speech, one group of 182 bits is passed through error correction and results in a string of 382 bits. The less critical group of 78 bits is just summed with the error-corrected data to yield 482 bits. These are diagonally interleaved to form 57 bits per time slot, the exact size of the data length in the data slots, two per burst. As a result, the 20-ms sample of speech is sent over eight timeslots over the air.

A further flag and section of 57 bits of data follow. Finally there are three final tail bits where the transmitter ramps down its power, and this is followed by a guard period to separate this burst from the next. The guard period is equivalent to 8.25 bits.

Vocoders

If digitized in a linear fashion, the speech would occupy a far greater bandwidth than any cellular system could accommodate. To overcome this, a variety of voice coding systems (or vocoders) are used. These systems involve analysing the incoming data that represent the speech and then performing a variety of actions upon the data to reduce the data rate. At the receiving end the reverse process is undertaken to re-constitute the speech data so that it can be understood.

The original vocoder used in GSM was called LPC-RPE (Linear Prediction Coding with Regular Pulse Excitation). This vocoder took each 20-ms block of speech and then represented it using just 260 bits. This actually equates to a data rate of 13 kbps.

In GSM, it is recognized that some bits are more important than others. If some bits are missed or corrupted, it is more important to the voice quality than others. Accordingly, the different bits are classified as:

- Class Ia, 50 bits – most important and sensitive to bit errors
- Class Ib, 132 bits – moderately sensitive to bit errors
- Class II, 78 bits – least sensitive to bit errors.

The 50 Class 1a bits are given a 3-bit Cyclic Redundancy Code (CRC) so that errors can be detected. This makes a total length of 53 bits. If there are any errors, the frame is not used and is discarded. In its place a version of the previously correctly received frame is used. These 53 bits, together with the 132 Class Ib bits with a 4-bit tail sequence, are entered into a 1/2 rate convolutional encoder. The total length is 189 bits. The encoder encodes each of the bits that enter as 2 bits, the output also being dependent upon a combination of the previous 4 input bits. As a result, the output from the convolutional encoder consists of 378 bits. The remaining 78 Class II bits are considered the least sensitive to errors, so they are not protected and are simply added to the data. In this way, every 20-ms speech sample generates a total of 456 bits. Accordingly, the overall bit rate is 22.8 kbps. Once in this format the data are interleaved to add further protection against interference and noise.

The 456 bits output by the convolutional encoder are divided into eight blocks of 57 bits, and these blocks are transmitted in eight consecutive time-slots – i.e. a total of four bursts, as each burst takes two sets of data (see Figure 6.4).

Later another vocoder called the Enhanced Full Rate (EFR) vocoder was added in response to the poor quality perceived by the users. This new vocoder gave much better sound quality, and was adopted by GSM. Using the ACELP (Algebraic Code Excitation Linear Prediction) compression technology, it gave a significant improvement in quality over the original LPC-RPE encoder. This became possible as the processing power that was available increased in mobile phones as a result of higher levels of processing power combined with lower current consumption.

There is also a half-rate vocoder. Although this gives far inferior voice quality, it does allow for an increase in network capacity. It is used in some instances when network loading is very high, to accommodate all the calls.

A further vocoder, known as Adaptive Multi-Rate (AMR), is a later development, and has provided some significant improvements in efficiency for networks – sometimes doubling the capacity. It is designed to work with both GSM full-rate (one user per each of eight time slots in each radio channel) and GSM half-rate (two users per time slot). AMR operates by using multiple voice encoding rates, each with a different level of error control. The AMR vocoder dynamically responds to radio conditions, using the most effective mode of operation at any given time. While the vocoder provides significant improvements in capacity it can also work under adverse conditions,

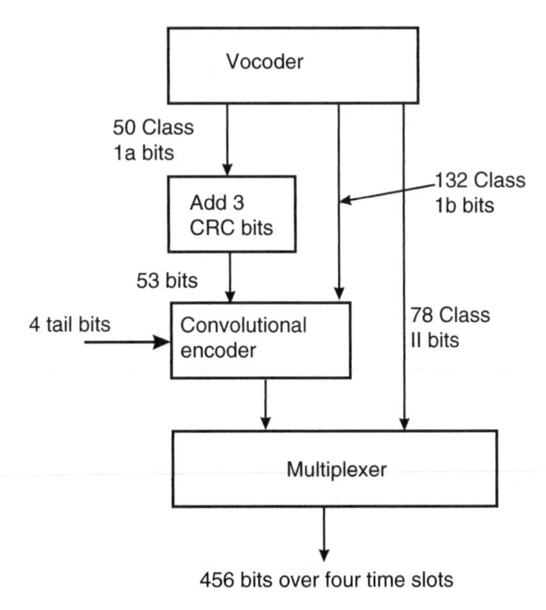

Figure 6.4 Vocoder set-up for GSM.

including poor radio signal conditions, where errors are high and when the network is heavily loaded. It is particularly useful at cell boundaries, where signal levels are lower and interference levels may be higher.

There are eight full-rate modes within AMR, and these operate at: 12.2, 10.2, 7.95, 7.4, 6.7, 5.9, 5.15 and 4.75 kbps. The highest rate of 12.2 kbps is actually the same as the GSM EFR vocoder and the 7.4 rate is the same as that used in ANSI-136 TDMA. The bit rate for each burst is 22.8 kbps. This includes both the voice data and the error correction information. By decreasing the voice data, the error correction can be increased to allow for increased errors and the system can therefore operate under poorer conditions by increasing the level of error correction. This can be achieved dynamically.

There are also six half-rate modes that operate at rates of 7.95, 7.4, 6.7, 5.9, 5.15 and 4.75 kbps. Since the gross bit rate in a half-rate channel is only 11.4 kbps, fewer bits are available for channel coding. Accordingly, a better signal-to-noise ratio is required for half-rate operation. However, when there is a good signal-to-noise ratio it can be enabled, allowing more users to access the same number of radio channels. AMR half-rate mode is further enhanced in EDGE radio networks, where more bits per time slot are available.

Operation

Initialization

When a GSM mobile phone is first turned on, it takes a few seconds before a call can be made. Not only does the mobile have to initialize itself, but it also needs to connect to the network so that it is ready to make and receive calls.

The first stage in linking to the network is to scan the available frequencies for a BTS. The mobile will scan all the available frequencies, noting the signals that can be received. It also checks for those signals that contain a control channel and notes the broadcast information, including the system identification and access control information, including the initial transmit power. The locations of messaging and paging channels are also sent. A unique identifier for the base station is sent, along with a network identifier.

The mobile chooses the strongest beacon frequency and then, using the random access channel, it sends a registration message. The network responds, checking the authentication and storing the location of the mobile in the HLR or VLR, as appropriate.

Call initiation

When a call is made from a mobile, the mobile has to communicate with the BTS so that the call can be set up and a connection made. It sends an access burst using the random access channel. This message contains a 5-bit number that temporarily identifies the mobile to the network. A further 3 bits contained within the message identify the type of message that is being requested – page response, call origination, or even a reconnection of a temporarily lost call.

There is a chance that another mobile will send a random access request at the same time, and if no acknowledgement is received it will wait a random amount of time so that it does not clash with the same mobile again, and then resend its access request.

When the BTS successfully receives the access request, it sends back the same random number and directs the mobile to a specific radio channel and time slot. Normally the network will request authentication and, if successful, the call can commence.

Receiving a call

A similar process to that used for a mobile-originated call is used for a call that needs to be received by a specific mobile. There are obviously some differences, because the mobile has to be alerted about the call rather than the mobile alerting the network that it needs to make a call.

When a request for a call is made to a specific mobile, the network checks the last registered location of the mobile with the location register. The network then sends a page message on the paging channel of a group of BTSs in the region where the mobile was last located.

When the mobile receives the paging message, it responds on its control channel. The network then allocates a radio channel and time slot to which the mobile is sent. Once there the call is made and continues until either end terminates the call.

Ending a call

A call may be terminated by either end of the connection. When this is detected, a number of messages are exchanged between the network and the mobile. Several messages are sent to ensure that the link has not been accidentally broken and that it is a real call termination.

Once the call is completed, the network frees the traffic channel for another user and the mobile returns to its idle state monitoring the paging channel. It also checks to ensure that the BTS it was using still offers the optimum signal. In this way it ensures that if any further calls arrive then it is in the best position to receive them.

Handovers

One of the key elements of a mobile phone is that it is able to move and still remain connected. This means that when the mobile moves out of the range of one BTS (i.e. out of a given cell) and into the next, it must be possible to transfer the call without any noticeable interruption to the user.

It takes a considerable amount of the network's time to ensure that this process, known as handover, happens correctly, as it is an essential element of the network control. Any problems quickly lead to customer dissatisfaction and a move to another network. The term for this is 'churn'.

In GSM which uses TDMA techniques the transmitter transmits only for one slot in eight and, similarly, the receiver receives only for one slot in eight. As a result, the RF section of the mobile could be idle for six slots out of the total eight. This is not the case, because during the slots in which it is not communicating with the BTS it scans the other radio channels looking for beacon frequencies that may be stronger or more suitable. In addition to this, when the mobile communicates with a particular BTS one of the responses it makes is to send out a list of the radio channels of the beacon frequencies of neighbouring BTSs. It scans these and reports back the quality of the link to the BTS. In this way the mobile assists in the handover decision, and as a result this is known as Mobile-Assisted HandOver (MAHO).

The network knows the quality of the link between the mobile and the BTS, as well as the strength of local BTSs as reported back by the mobile. It also knows the availability of channels in the nearby cells. As a result it has all the information it requires to be able to make a decision about whether it needs to hand the mobile over from one BTS to another.

If the network decides that it is necessary for the mobile to hand over, it assigns a new channel and time slot to the mobile. It informs the BTS and the mobile of the change. The mobile then retunes during the period it is not transmitting or receiving, i.e. in an idle period.

Frequency hopping

One of the facilities that the GSM standard offers is to enable a mode of operation called frequency hopping. This is effectively a form of spread spectrum operation. Essentially, when a signal uses frequency hopping it moves from one frequency to another, staying on a particular frequency for a short length of time, sufficient to send a burst of data. It then moves onto another frequency, where it sends another burst of data. To operate in this mode, both transmitter and receiver must follow the same hop pattern so that they are both on the same frequency at the same time.

There are a number of advantages to frequency hopping. It is a technique that is used by the military to prevent jamming by hostile signal sources, and it also prevents eavesdropping because the signal cannot be received easily unless the hop pattern is known. For applications such as cellular telecommunications, it is primarily used because it averages the level of interference. By using frequency hopping, if one channel is blocked this will have only a transitory effect. Also, as the channel is effectively shared by several mobiles hopping from one channel to the next, this averages the level of interference. This has significant advantages for network planners, who often have to design systems for worst-case situations; by adopting frequency hopping, average scenarios can be taken. As a result of this, the level of frequency re-use can be improved, thereby bringing considerable operational and financial advantages.

A further advantage is that it reduces the effects of selective fading. A signal that reaches a mobile or BTS will be the sum of signals from the transmitter reaching the receiver via several paths as a result of reflections. As the lengths of the paths will differ, sometimes the signals will add constructively to produce a larger signal, whereas at other times they will tend to cancel each other out. As this is somewhat dependent on the frequency in use, moving to a different channel will often ease the situation. Again, the problem tends to be averaged out.

Frequency hopping is relatively easy to implement in GSM because the RF section of the mobile is already moving between different frequencies for transmit and receive, changing frequency during the dead periods when neither the transmitter nor receiver are active.

In order to coordinate the transmit and receive channels and ensure that the BTS and mobile hop in synchronism, the hopping algorithm is broadcast on the Broadcast Control Channel.

Short message service

The Short Message Service (SMS) is a feature that is widely used on GSM mobiles today. It provides the ability to send and receive text messages to and from mobile telephones. The text can comprise words or numbers, or an alphanumeric combination. SMS was created as part of the GSM Phase 1 standard, and its use has grown beyond all expectations. The first short message is believed to have been sent in December 1992 from a Personal Computer (PC) to a mobile phone on the Vodafone GSM network in the UK. Each short message is up to 160 characters in length when Latin alphabets are used, and 70 characters in length when non-Latin alphabets (such as Arabic and Chinese) are used.

SMS is what is termed a 'store and forward' service. The messages from the sender are routed to an SMS centre and then on to the recipient. Each mobile telephone network that supports SMS has one or more messaging centres to handle and manage the short messages. Once delivered a confirmation of message is delivered, and this can be selected to appear at the user's handset.

Short messages can be sent and received simultaneously with GSM voice, data and fax calls. This is possible because voice, data and fax calls take over the traffic channel, whereas the SMS uses the signalling path.

General packet radio service

As mobile phone technology became more widely used, it became obvious that data services would become the next major generator of revenue. Although many second-generation systems were able to carry data, it was only at very low data rates. For GSM, the maximum rate was 14.4 kbps.

As a result of this, developments were made to enable higher speed data to be carried over the networks. The first was known as High Speed Circuit Switched Data (HSCSD). Although this was never widely deployed, it enabled data rates of up to about 64 kbps to be achieved by combining time slots. Although this enabled higher-speed data to be transferred, the concept was not widely taken up because it did not make efficient use of the resources of the network.

Most data transfer occurs in what is often termed a 'bursty' fashion. The transfer occurs in short peaks, and then remains quiescent for a while. This means that the channel remains dormant, even though this capacity could be used to transfer data for other users. To overcome this problem, a system called the General Packet Radio Service (GPRS) was devised. Rather than using a circuit switched scheme, where a complete circuit is devoted to a user, this scheme uses packet data and enables a far greater level of efficiency to be obtained.

In addition to this, the data world is now very focused on the Internet and the protocols that it has adopted. The 3G services have an IP-based structure, and with this in place it is easier to migrate to the 3G standards such as UMTS/W-CDMA.

GPRS network structure

In order for the GPRS network structure to provide a packet-based service, the core network structure has to be upgraded from what is used for GSM. The network between the BSC and BTS is similar, but behind this, new infrastructure is required to support the packet data.

Data from the BSC is routed through what is termed a Serving GPRS Support Node (SGSN), which forms the gateway to the services within the network, and then a Gateway GPRS Support Node (GGSN), which forms the gateway to the outside world. These elements form what is known as the Public Land Mobile Network (PLMN).

The SGSN serves a number of functions for mobiles with a data service. It enables authentication to occur, and it then tracks the location of the mobile within the network and ensures that the quality of service is of the required level.

For the network protocols there are two layers that are used and supported by GPRS, namely X25 and IP. In operation, the protocols assign addresses (Packet Data Protocol or PDP addresses) to the devices in the network for the purpose of routing the data through the system. Thus the GGSN appears as a data gateway to the public packet network, and therefore the fact that the users are mobiles cannot be identified.

In operation, the mobile must attach itself to the SGSN and activate its PDP address. This address is supplied by the GGSN, which is associated with the SGSN. As a result a mobile can attach to only one SGSN although, once assigned its address, it can receive data from multiple GGSNs using multiple PDP addresses.

Layers

Software plays a very large part in the current cellular communications systems. To enable it to be sectioned into areas that can be addressed separately, the concept of 'layers' has been developed. It is used in GSM and other cellular systems, but as they become more datacentric the idea takes a greater prominence. Often these are referred to as layers, 1, 2 and 3.

Layer 1 concerns the physical link between the mobile and the base station. This is often subdivided into two sub-layers, namely the physical RF layer, which includes the modulation and demodulation, and the physical link layer, which manages the responses and controls required for the operation of the RF link. These include elements such as error correction, interleaving and the correct assembly of the data, power control, and the like.

Above this are the Radio Link Control (RLC) and the Medium Access Control (MAC) layers. These organize the logical links between the mobile and the base station. They control the radio link access and organize the logical channels that route the data to and from the mobile.

There is also the Logical Link Layer (LLC), which formats the data frames and is used to link the elements of the core network to the mobile.

GPRS mobiles

In view of the increased capabilities that GPRS offers, the mobiles that are available are classified according to their capabilities in terms of both the ability to connect to GSM and GPRS facilities, and also the data rates they support. In the first instance, they are classified into three classes to describe the way in which they can connect to the services available:

1. *Class A* describes mobile phones that can be connected to both GPRS and GSM services at the same time.
2. *Class B* mobiles can be attached to both GPRS and GSM services, but they can be used on only one service at a time. A Class B mobile can make or receive a voice call, or send and or receive a SMS message during a GPRS connection. During voice calls or texting the GPRS service is suspended, but it is re-established when the voice call or SMS session is complete.
3. *Class C* phones can be attached to either GPRS or GSM services, but the user needs to switch manually between the two different types.

The mobiles are also categorized by the data rates they can support. In GSM there are eight time slots that can be used to provide TDMA, allowing multiple mobiles onto a single RF signal carrier. With GPRS it is possible to use more than one slot to enable much higher data rates to be achieved when these are available. The different speed classes of the mobiles are dependent upon the number of slots that can be used in either direction. There are a total of 29 speed classes: class 1 mobiles are able to send and receive in one slot in either direction (i.e. uplink and downlink), and class 29 mobiles are able to send and receive in all eight slots. The classes between these two limits are able to support sending and receiving in different combinations of uplink and downlink slots.

GPRS coding

GPRS offers a number of coding schemes with different levels of error detection and correction. Their use is dependent upon the radio-frequency signal conditions and the requirements for the data being sent. These are given labels CS-1 to CS-4:

- *CS-1* applies the highest level of error detection and correction, and is used in scenarios when interference levels are high or signal levels are low. By applying high levels of detection and correction, this prevents the data having to be re-sent too often. Although it is acceptable for many types of data to be delayed, for others there is a more critical time element. This level of detection and coding results in a half code rate – that is, for every 12 bits that enter the coder, 24 bits result. It results in a 9.05-kbps actual throughput data rate.

- *CS-2* is an error detection and coding scheme for better channels. It effectively uses a 2/3 encoder and results in a real data throughput of 13.4 kbps, which includes the RLC/MAC header etc.
- *CS-3* effectively uses a 3/4 coder and results in a data throughput of 15.6 kbps.
- *CS-4* is used when the signal is high and interference levels are low. No correction is applied to the signal, allowing for a maximum throughput of 21.4 kbps. If all eight slots were used, this would enable a data throughput of 171.2 kbps to be achieved.

In addition to the error detection and coding schemes, GPRS also employs interleaving techniques to ensure the effects of interference and spurious noise are reduced to a minimum. It allows the error correction techniques to be more effective, as interleaving helps to reduce the total corruption if a section of data is lost.

As 20-ms blocks of data are carried over four bursts, with a total of 456 bits of information, a total of 181, 268, 312 or 428 bits of payload data are carried, dependent upon the error detection and coding scheme chosen – i.e. from CS-1 to CS-4, respectively.

GPRS physical channel

GPRS builds on the basic GSM structure. It uses the same modulation and frame structure that is employed by GSM, and in this way it is an evolution of the GSM standard. Slots can be assigned dynamically by the BSC to GPRS, dependent upon the demand, the remaining ones being used for GSM traffic.

There is a new data channel that is used, called the Packet Data Channel (PDCH). The overall slot structure for this channel is the same as that used within GSM, having the same power profile and timing advance attributes to overcome the different signal travel times to the base station dependent upon the distance the mobile is from the base station. This enables the burst to fit in seamlessly with the existing GSM structure.

Each burst of information is 0.577 ms in length, and is the same as that used in GSM. It also carries two blocks of 57 bits of information, giving a total of 114 bits per burst. It therefore requires four bursts to carry each 20-m block of data, i.e. 456 bits of encoded data (Figure 6.5).

The BSC assigns PDCHs to particular time slots, and there will be times when the PDCH is inactive, allowing the mobile to check for other base stations and monitor their signal strengths

Tail bits	Encrypted data	Training sequence	Encrypted data	Tail bits	Guard period
3 bits	57 bits + 1-bit coding scheme flag	26 bits	57 bits + 1-bit coding scheme flag	3 bits	8.25 bits

Figure 6.5 Physical structure of a GPRS burst.

to enable the network to judge when handover is required. The GPRS slot may also be used by the base station to judge the time delay, using a logical channel known as the Packet Timing Advance Control Channel (PTCCT).

Channel allocation

Although GPRS uses only one physical channel (PDCH) for the sending of data, it employs several logical channels that are mapped into this to enable the GPRS data and facilities to be managed. As the data in GPRS are handled as packet data rather than circuit switched data, the way in which this is organized is very different to that on a standard GSM link. Packets of data are assigned a space within the system according to the current needs, and routed accordingly.

The MAC layer is central to this, and there are three MAC modes that are used to control the transmissions: fixed allocation, dynamic allocation and extended dynamic allocation.

The fixed allocation mode is required when a mobile requires data to be sent at a consistent data rate. To achieve this, a set of PDCHs are allocated for a given amount of time. When this mode is used there is no requirement to monitor for availability, and the mobile can send and receive data freely. This mode is used for applications such as video-conferencing.

When using the dynamic allocation mode, the network allocates time slots as they are required. A mobile is allowed to transmit in the uplink when it sees an identifier flag known as the Uplink Status Flag (USF) that matches its own. The mobile then transmits its data in the allocated slot. This is required because up to eight mobiles can have potential access to a slot, but obviously only one can transmit at any given time.

Extended dynamic allocation is also available. Use of this mode allows much higher data rates to be achieved because it enables mobiles to transmit in more than one slot. When the USF indicates that a mobile can use this mode, it can transmit in the number allowed, thereby increasing the rate at which it can send data.

Logical channels

There is a variety of channels used within GPRS, and they can be set into groups dependent upon whether they are for common or dedicated use (see below). Naturally, the system does use the GSM control and broadcast channels for initial set-up, but all the GPRS actions are carried out within the GPRS logical channels carried within the PDCH.

1. *Common control channels*

   - Packet Broadcast Control Channel (PBCCH). This is a downlink-only channel that is used to broadcast information to mobiles and informs them of incoming calls etc. It is very similar in operation to the BCCH used for GSM. In fact the BCCH is still required in the initialization to provide a time slot number for the PBCCH. In operation, the PBCCH broadcasts general information such as power control parameters, access methods and operational modes, network parameters, etc., required to set up calls.

- Packet Paging Channel (PPCH). This is a downlink-only channel, and is used to alert the mobile to an incoming call and to warn it to be ready to receive data. It is used for control signalling prior to the call set-up. Once the call is in progress, a dedicated channel referred to as the PACCH takes over.
- Packet Access Grant Channel (PAGCH). This is also a downlink channel, and it sends information telling the mobile which traffic channel has been assigned to it. It occurs after the PPCH has informed the mobile that there is an incoming call.
- Packet Notification Channel (PNCH). This is another downlink-only channel which is used to alert mobiles that there is broadcast traffic intended for a large number of mobiles. It is typically used in what is termed 'point-to-point multicasting'.
- Packet Random Access Channel (PRACH). This is an uplink channel that enables the mobile to initiate a burst of data in the uplink. There are two types of PRACH burst; one is an 8-bit standard burst, and a second one using an 11-bit burst has added data to allow for priority setting. Both types of burst allow for timing advance setting.

2. *Dedicated control channels*

- Packet Associated Control Channel (PACCH). This channel is present in both uplink and downlink directions, and is used for control signalling while a call is in progress. It takes over from the PPCH once the call is set up, and it carries information such as channel assignments, power control messages and acknowledgements of received data.
- Packet Timing Advance Common Control Channel (PTCCH). This channel, which is present in both the uplink and downlink directions, is used to adjust the timing advance. This is required to ensure that messages arrive at the correct time at the base station, regardless of the distance of the mobile from the base station. As timing is critical in a TDMA system and signals take a small but finite time to travel, this aspect is very important if long guard bands are not to be left.

3. *Dedicated traffic channel*

- Packet Data Traffic Channel (PDTCH). This channel is used to send the traffic and it is present in both the uplink and downlink directions. Up to eight PDTCHs can be allocated to a mobile to provide high-speed data.

GPRS operation

There are three modes in which a GPRS mobile may operate: initialization/idle, standby and ready.

When the mobile is turned on, it must register with the network and update the location register. This is very similar to the action performed with a GSM mobile, but it is referred to as a location update. It first locates a suitable cell and transmits a radio burst on the RACH, using a shortened burst because it does not know what timing advance is required. The data contained within this burst temporarily identifies the mobile, and indicates that the reason for the update is to perform a location update.

When the mobile performs its location update, the network also performs an authentication to ensure that it is allowed to access the network. As with GSM, it accesses the HLR and VLR as necessary for the location update and the AuC for authentication. It is at registration that the network detects that the mobile has a GPRS capability. The SGSN also maintains a record of the location of the mobile so that data can be sent there if required.

The mobile then enters a standby mode, periodically updating its position. It monitors the MNC of the base station to ensure that it has not changed base stations, and looks for stronger base-station control channels.

The mobile also monitors the PPCH in case of an incoming alert indicating that data are ready to be sent. As with GSM, most base stations set up a schedule for paging alerts based on the last figures of the mobile number. In this way it does not have to monitor all the available alert slots, and can instead monitor a reduced number to which it knows alerts can be sent for it. In this way the receiver can be turned off for longer and battery life can be extended.

In the ready mode, the mobile is attached to the system and a virtual connection is made with the SGSN and GGSN. By making this connection, the network knows where to route the packets when they are sent and received. In addition to this, the mobile is likely to use the PTCCH to ensure that its timing is correctly set so that it is ready for a data transfer should one be needed.

With the mobile attached to the network, it is prepared for a call or data transfer. To transmit data, the mobile attempts a Packet Channel Request using the PRACH uplink channel. As this may be busy, the mobile monitors the PCCCH, which contains a status bit indicating the status of the base station receiver – whether it is busy, or idle and capable of receiving data. When the mobile sees this status bit indicating that the receiver is idle, it sends its packet channel request message. If accepted, the base station will respond by sending an assignment message on the PAGCH on the downlink. This will indicate which channel the mobile is to use for its packet data transfer, as well as other details required for the data transfer.

This sets up the packet data transfers for the uplink only. If data need to be transferred in the downlink direction, then a separate assignment is performed for the downlink channel.

When data are transferred, this is controlled by the action of the MAC layer. In most instances it will operate in an acknowledge mode, whereby the base station acknowledges each block of data. The acknowledgement may be contained within the data packets being sent in the downlink, or the base station may send data packets down purely to acknowledge the data.

When disconnecting, the mobile sends a packet temporary block flow message, and this is acknowledged. Once this has taken place, the USF assigned to the mobile becomes redundant and can be assigned to another mobile wanting access. At this point the mobile effectively becomes disconnected and, although still attached to the network, no more data transfer takes place unless it is re-initiated. Separate messages are needed to detach the mobile from the network.

EDGE

EDGE is a further evolution of GSM, enabling higher data rates to be achieved but using substantially the same network equipment. Effectively, it is an upgrade to GPRS. Although upgrades are required at the base station, as there are differences to the radio interface, the core network capability remains virtually unchanged, allowing migration to the new standard. EDGE stands for either Enhanced Data for Global Evolution or for Enhanced Data for GSM Evolution – both are widely used and accepted.

One of the main differences between EDGE and GPRS is that EDGE uses a different form of modulation. GPRS uses the GMSK used on GSM, whereas EDGE, to achieve higher data rates, uses a higher-order modulation format in the form of 8PSK (see Chapter 3). This means that instead of the maximum payload per slot being 116 bits, as in the case of GPRS, it rises to 464 bits. The packet technology used in GPRS is carried forward to EDGE, and this means that the upgrades that are required are essentially within the BTS to enable it to transmit and receive 8PSK modulation.

EDGE offers a significant speed improvement over GPRS, but this can only be achieved under ideal conditions. The optimum radio conditions occur only when the signals in both directions are sufficiently strong, and this normally entails the mobile being reasonably close to the base station. Once a mobile starts to move, variations in signal strength are normally experienced and these degrade the quality of the link.

Although 8PSK, being a higher-order modulation format, enables a three-fold increase in the data transfer rate, this comes at the price of noise resilience. It is found that 8PSK is considerably more sensitive to noise than GMSK, and this is why the higher data rate is available only under good signal conditions.

Phones equipped for EDGE can be used for both circuit switched voice and packet switched data. They can also operate on GPRS-only networks. In fact, the phone will use the higher-order 8PSK modulation only when the conditions are suitable, returning to the GMSK modulation when the link is not optimum for 8PSK.

The phones are grouped into classes dependent upon their capabilities. Class A equipment is able to support simultaneous GSM voice along with GPRS/EDGE data; class B equipment can support GSM voice and GPRS/EDGE data, but not simultaneously; and class C equipment can support only GPRS/EDGE data services but no voice services.

Time slots

EDGE, GPRS and GSM all have to operate alongside each other in a network. It is a primary requirement that evolutionary technologies are all able to operate on the same network. This ensures that service is offered to existing customers using older phones along with those paying additional rates for the premium EDGE services. The network therefore has to support both services operating simultaneously. Accordingly, different slots within the traffic frames will need to be able to support different structures and different types of modulation, dependent upon the

Table 6.2 Comparison of GPRS and EDGE data payloads.

	GPRS (using GMSK)	EDGE (using 8PSK)
Overall payload per slot	116	464
Overall payload per block	348	1392

phones being used, the calls being made and the prevailing conditions. It is quite possible that one slot may be supporting a GSM call, the next a GPRS data connection and the third an EDGE connection using 8PSK.

Table 6.2 provides a comparison of GPRS and EDGE data payloads.

Data coding and throughput

Most of the data being sent over an EDGE link consist of TCP/IP packets. These packets are longer than a single EDGE packet payload, and therefore it is necessary to split the TCP/IP packets into smaller sections known as 'chunks'. These chunks have defined sizes, and may consist of one of 22, 28, 34 or 37 bytes or 'octets'. The 37-octet chunk may be made directly of data to be transmitted, or it may be a 34-octet chunk which is then padded by adding three dummy octets.

There are nine different Modulation and Coding Schemes (MCS) that can be used with EDGE, and each one is designated a number in the region 1 to 9. These allow different degrees of error protection (and coding rate), resulting in a change in the net data throughput. The system detects the number of bit errors and adjusts the coding scheme accordingly. It naturally endeavours to adopt the scheme that will result in the highest throughput, but will adjust itself according to the prevailing conditions, changing as required.

The different coding schemes are grouped into three classes or families, which are referred to as families A, B and C. The coding schemes within a family are used together and complement one another (Table 6.3). Family A consists of MCS-3, MCS-6, MCS-8 and MCS-9; family B consists

Table 6.3 EDGE error protection and coding schemes.

Scheme name	Effective code rate	Modulation format	Data rate using one time slot (kbps)
MCS-9	1.0	8PSK	59.2
MCS-8	0.92	8PSK	54.4
MCS-7	0.76	8PSK	44.8
MCS-6	0.49	8PSK	29.6
MCS-5	0.37	8PSK	22.4
MCS-4	1.0	GMSK	17.6
MCS-3	0.8	GMSK	14.8
MCS-2	0.66	GMSK	11.2
MCS-1	0.53	GMSK	8.8

of MCS-2, MCS-5 and MCS-7; and family C consists of MCS-1 and MCS-4. The advantage of grouping the families together in this way is that if a block transmitted in one of the coding schemes is not acknowledged, then it can be sent as two blocks with a coding scheme in the same family. For example, if a block transmitted using MCS-7 is corrupted, then it can be re-sent as two blocks using MCS-5 or four using MCS-2.

Operation

The overall operation of an EDGE network is virtually identical to that of GPRS. The same core network components are used, as it is a migration from one to the next allowing higher data rates. In fact, the differences occur at the air interface. Here, the new scheme sends data faster and is more adaptive in terms of the modulation scheme and coding schemes used, so that the optimum performance is achieved. It is also possible for a network to run simultaneous GSM, GPRS and EDGE calls. The major upgrades are to the base station, where new hardware is normally required to enable the transmission of 8PSK as well as GMSK modulation.

North American TDMA

While GSM was the second-generation digital cellular system that was introduced within Europe initially, another development took place in the USA. There were different requirements, and as a result the system, although digital and using Time-Division Multiple Access (TDMA) techniques, is very different.

The FCC had indicated that no new frequency allocations were to be made to accommodate a new cellular system. Therefore, with a very heavy investment in AMPS technology which had been deployed over vast areas of the USA, there was a need to ensure that any digital system that was developed was as compatible as possible with AMPS. In view of this, the system was designed to use the existing AMPS channels so that it would allow a smooth transition from AMPS to TDMA. During the changeover, both AMPS and TDMA could co-exist in the same area.

Although using the same 30-kHz channel spacing as adopted by AMPS, TDMA provides a threefold increase in voice traffic by splitting the voice channel using digitized voice, and using time slots for three voice channels. In addition to the improvement in capacity, the system has also been made secure against casual listening, cloning and other forms of fraud.

The system, which is primarily used only within North America, is known by a variety of names. Today it is most commonly called just TDMA, in view of its time-division multiple access technology. However, outside the USA it is often called NA or North American TDMA, for obvious reasons. It is also called Digital AMPS (DAMPS) as well as US Digital Cellular (USDC), American Digital Cellular (ADC) and North American Digital Cellular (NADC). Whatever name is used, it is the same system.

The system was originally defined under the standard IS-54. However, the performance of the system was upgraded to keep pace with some of the competing technologies, and IS-54 was superseded by IS-136. This includes new features such as text messaging and data capabilities.

Although the system in this format is used within North America, a variant was adopted in Japan. Known as Pacific or Personal Digital Cellular (PDC), this is used exclusively in Japan, where has it attained a very high degree of use and market penetration.

System overview

The NA-TDMA standard builds on many of the elements found in AMPS, containing many of the features used in AMPS but implemented in a format suitable for a digital standard. It utilizes the idea of control channels, as well as using different traffic channels onto which the calls are routed. There are also features that are similar to the SAT tones used in AMPS.

The overall network architecture is similar to that of AMPS. Base stations are grouped together under the control of a base station controller. This will manage or control a small number of base stations and enable some of the local decisions and control to be effected close to the area where the control is needed. This simplifies the central control that is required, and also makes communication faster and more effective. The base station controllers are then connected back to the mobile switching centre, which takes overall control of the network and links into the PSTN. As before, the MSC contains the location registers for monitoring the location of mobiles on the network, and also the database used for authentication of the mobiles attempting to gain access to the network.

RF signal

The signals that are transmitted in NA-TDMA fall within a 30-kHz channel spacing, as already discussed. To achieve this spacing while being able to accommodate a sufficiently high data rate to support three digitized voice channels, a form of modulation known as DQPSK is used. This enables data at a rate of 48.6 kbps to be fitted into the channel bandwidth of 30 kHz.

This approach provides a very high level of modulation efficiency. The overall level is 1.62 bps/Hz (i.e. 48.6 kbps/30 kHz). Unfortunately there is a price to pay for this. One of the major disadvantages is that the signal varies in amplitude as the modulation is applied. GSM uses a form of modulation known as GMSK, and the amplitude of this remains constant. The use of a form of modulation that gives a varying output amplitude requires that a linear amplifier is used in the final stages of the transmitter of the mobile. As linear amplifiers are less efficient than those that are non-linear, this reduces battery life. However, the system requires a very high modulation efficiency, and this was part of the compromise needed in choosing the most suitable form of modulation.

Even though NA-TDMA operates on 30-kHz channels, the digital information causes the spectrum of the radio frequency signal to be different to that of the FM signal modulated with straight analogue audio. As a result, the mask used to define the maximum bandwidth of the digital signal is less stringent than that of the analogue transmissions, although still retaining a sufficiently low level of signal in the adjacent channels to enable interference to be kept within acceptable limits.

In addition to the 850-MHz band that was used for AMPS and the first deployment of NA-TDMA defined under IS-54, the later release of the standard released under IS-136 allows for transmissions in the PCS band at 1900 MHz. This has given further capacity to systems that were reaching their limits. As a result, the frequency bands 869–894 MHz and 1930–1990 MHz are used for the base station transmissions, and 824–849 MHz and 1850–1910 MHz for the mobile transmissions.

Channels

When the first forms of NA-TDMA as defined under IS-54 were deployed the system operated in a very similar way to AMPS, but instead of having an analogue voice or traffic channel it was digital and enabled three voice channels to be accommodated on the single RF carrier. By allowing the AMPS FOCC and RECC channels to remain, it was possible to run the two systems together and give backward compatibility. Nevertheless, some upgrades were required to enable the additional facilities required by NA-TDMA to operate.

In total, there are four messages required for authentication, three for FOCC and two for RECC. These provide the required level of security to prevent a variety of forms of fraud that were prevalent when using AMPS, such as cloning.

The message structure using RECC and FOCC has been updated to provide more flexibility and allow both voice and data to be transferred, thereby allowing data (including SMS) to be carried. On RECC, two additional messages have been introduced; Page Response With Service, and Origination With Service. On FOCC, Page With Service and Message Waiting have been introduced.

In order to direct the mobile to the correct digital channel for the call, a message known as Initial Digital Traffic Channel has been introduced onto the FOCC. This enables the system to use the FOCC and divide traffic channels between analogue and digital, thereby allowing analogue and digital formats to operate satisfactorily alongside each other.

A further addition is new information that has been added into the Global Action Message. The basic message was used in AMPS, but by adding a DCCH (Digital Control Channel) pointer it enables the mobile to move to the allocated DCCH to receive its service.

Identifiers

Although the basic channel structure is approximately the same, there are naturally some differences. There are both control and traffic channels. However, the base stations transmit what is called a Digital Verification Colour Code (DVCC). This performs a very similar function to that of the SAT signal transmitted for AMPS, where the tone indicated the group of stations to which a particular base station belonged. Using the DVCC, a specific digital identification known as a LOCAID is transmitted. This is allocated to a group of base stations, and provides a broad indication of the area in which a mobile is located. If the mobile moves into a different area, it sees a new LOCAID. By knowing the LOCAID the mobile is receiving, the network knows which base stations need to page the mobile for an incoming call. This considerably reduces the load on the network as it only has to send the page message to a limited number of base stations.

Also, by sending the page to a number of base stations in the vicinity of the mobile, it is possible to ensure that it receives the message even if it has moved slightly.

Another identifier, known as the System Operator Code (SOC), identifies the carrier or network owner. This is transmitted by the base stations so that mobiles know the network to which they are connecting. This is particularly important for roaming.

A further new identifier that has been added is what is called the protocol version. With software being continually upgraded, this is used to indicate the capabilities of the mobile in question.

Finally, a 64-bit authentication key is assigned by the carrier. Although not transmitted, this key is used for authentication against possible fraud. It helps to overcome some of the problems that were encountered with the analogue systems and, in this case, AMPS.

Digital traffic channel

Obviously one of the important and highest profile channels is the Digital Traffic Channel (DTCH). This carries the digitized voice and other payload required. The channel is divided into frames, each of 40 ms (i.e. twenty-five frames are transmitted every second). These frames are further divided into six slots, each slot being 6.67 ms long. Two slots are reserved for each user: 1 and 4, 2 and 5, and 3 and 6 form the pairs.

In terms of the bits carried, the overall frame consists of 1944 bits or 972 symbols, as the pi/4 DQPSK used carries 2 bits per symbol. Each of the slots carries 324 bits or 162 symbols.

Data within each slot are organized into a number of logical channels or fields. The format for the slots is different between the uplink and the downlink, as shown in Figure 7.1.

A variety of channels or fields can be seen in the diagram. When looking at the channels in the forward and reverse links, it should be remembered that the forward link (from the base station to the mobiles) transmits continually whereas the reverse link (from the mobile to the base station)

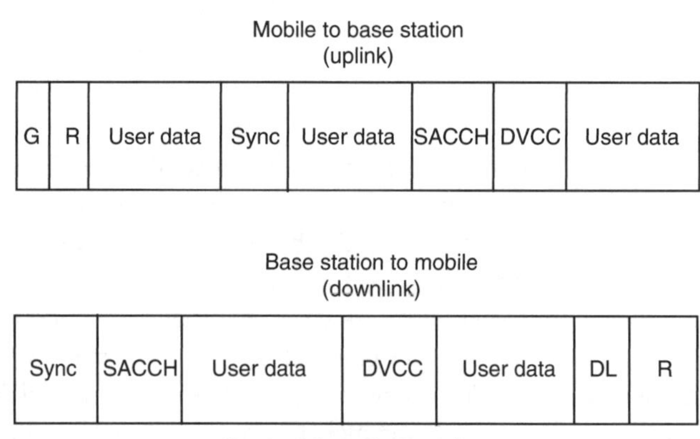

Figure 7.1 Logical channels or fields within the Data Traffic Channel.

transmits only during the allocated slots. This is part of the reason for the difference in structure between the slots being transmitted in either direction.

For the mobile to transmit to the base station, the first section of the slot is a guard band. This serves to allow the transmitter to turn on, and also no data are transmitted in this period as there may still be some residual interference from the mobile just finishing its slot. Part of the reason for this is the varying distances of the mobiles from the base station, and the propagation time for the signal. The time reserved for this has a time corresponding to 6 bits. The next element of time, again 6 bits long, is reserved for ramp-up time, to allow the transmitter in the mobile to turn on and come up to full power.

The sync field obviously enables synchronization to be effected. The process known as time alignment is critical in a TDMA system, where different mobiles are at different distances from the base station. Using this information, the mobile is able to advance its transmission time correctly to ensure that its transmission fits within the allocated slot at the base station. The sync field is also used in a process termed adaptive equalization, which allows the receiver to compensate for differences between the actual received waveform and what is normally expected.

To align the mobile completely with the base station timing, it is sometimes necessary to undertake a more rigorous alignment process. In order that the transmitted burst does not fall outside the allotted timeslot, even with the delays, this burst is made shorter. It uses a guard bit field of 50 bits rather than the normal 6 bits, and then only sync and DVCC (digital verification colour code; see below). This is transmitted repeatedly until the base station is able to calculate the required time alignment value.

A further field is known as the Digital Verification Colour Code (DVCC). This is a development of the SAT signal used on AMPS, and is used to ensure that the mobile is communicating correctly with the base station. To enable a mobile to be accepted by a base station, it must correctly receive the DVCC data and retransmit it back to the base station. Only when the data transmitted and received by the base station are the same does the base station give access to the mobile.

Another field is known as the Slow Associated Control Channel (SACCH). This allows control data messaging to be undertaken without interfering with the voice traffic. The message format for the SACCH is 132 bits long, but as it is possible to send only 12 bits per slot this means that it takes 11 slots (or six frames) to deliver the complete message.

The data rate allowed by the SACCH is sufficiently high for many of the control messages that need to be sent. However, for operations such as handoff a faster rate is required. To achieve this, a further channel known as the Fast Associated Control Channel (FACCH) is used. This occupies the same slots as the voice or traffic data being transmitted, displacing it as required. In this way it is able to use up to 260 bits per slot. To overcome the problem of the displaced traffic data a scheme known as bad-frame masking is used, whereby the receiver re-uses the last blocks received to reduce the effect of the missing traffic data.

A final control channel is known as the Digital Control Channel Locator (DL). This is used to inform the mobile which control channel to move to once the call has terminated.

The last field of channel is the one that is the focus of the call, namely the data channel or data field that is used to carry the voice traffic data. A vocoder known as ACELP (Algebraic Code Excited Linear Prediction) is used. The voice is digitized and compressed, and split into 20-ms blocks of data, each 260 bits long. This represents a data rate of 13 kbps.

The messages that are carried on the DTCH are contained within either the SACCH or the FACCH, and they are very similar in structure to those used on AMPS.

Digital control channel

The Digital Control Channel (DCCH) is used to control the mobile and pass information to it to enable calls to be set up and controlled. The top-level structure consists of hyperframes lasting 1.28 seconds. Within these hyperframes, two superframes are contained. These are 0.64 seconds long, and they themselves contain sixteen frames, each having a duration of 40 ms. The frames are then split up into two blocks of 20 ms, and in turn these are split into three slots. When in use, two slots are used per frame for full-rate transmissions while only one slot is used for half-rate transmissions (Figure 7.2).

The channel is set up with a number of logical channels or fields. Sync is used to ensure that the correct timing is achieved, and for this channel an additional preamble is used in the reverse link. For the forward link, a SuperFrame Phase (SFP) indicator is used to indicate the current block being transmitted within the superframe. This provides two functions: first, timing information; and secondly, it enables the mobile to examine the structure and determine whether it is receiving DCCH or DVCH.

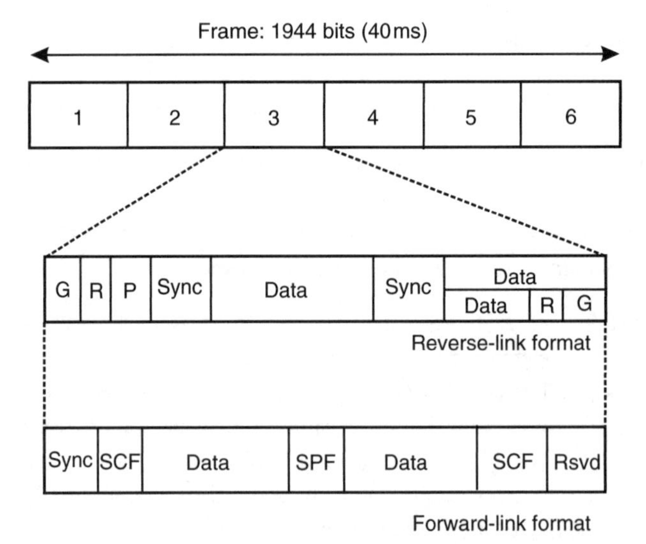

Figure 7.2 DCCH structure.

A field referred to as the Shared Channel Feedback (SCF) field is used to maintain the reverse channel transmissions. It contains three sub-fields, namely the BRI, R/N and CPE fields. The first field is the Busy/Reserved/Idle (BRI) field, which informs the mobile whether the RACH is being used, and hence whether it can access the system to make a call. The Received/Not Received (R/N) field is used to indicate the result of a random access, along with the Coded Partial Echo (CPE) field. Additionally, the user data may be abbreviated in a shortened burst and, as shown in Figure 7.2, one of two formats may be adopted dependent upon the amount of data to be transmitted.

Data on the forward channel falls into two categories or channels, namely a BroadCast CHannel (BCCH) and a Short message Paging and Access response CHannel (SPACH).

Paging

The way in which mobiles are paged can have a bearing on a number of elements of the overall performance. One in particular is the battery life. By arranging the paging messaging such that many of the circuits within the mobile can be switched off while the mobile is not involved in a call, it is possible to conserve power and considerably extend the life of the battery. Accordingly, mobiles have a 'sleep' mode and the channel structure has been organized to facilitate this.

When there is an incoming call for a mobile, the mobile is alerted using the paging sub-channels in the SPACH blocks. These are transmitted twice, once in each superframe found in the hyperframes. By enabling the mobile to close down during some superframes and listen in to specific ones only, it is able to shut off its circuitry for part of the time.

The MIN within the mobile determines which paging channels the mobile will listen to, and the length of time the mobile is allowed to sleep is determined by the Paging Frame Number. A number 1 means that the mobile will monitor one paging channel in each hyperframe. The maximum 96 means that the mobile will sleep through 96 hyperframes. While the mobile sleeps, it cannot pick up paging messages. As each hyperframe takes 1.28 seconds, if the PFN is equal to 1 then the mobile will be able to identify an incoming call quickly. A PFN of 96 equates to 122.88 seconds, meaning that it might take up to this time for the mobile to realize that there is an incoming call. However, battery life in the idle mode would be very good.

Handoff

In order to determine whether the station needs to handoff to another base station, the mobile makes a determination of the bit error rate (BER). This gives not only an indication of the signal strength, but also the level of interference. This is a far more useful indication of whether a handoff is required.

The mobile is provided with information detailing channels where it might look for other local base stations. This information is contained within a message called the Measurement Order, and it contains six or twelve different channels where the mobile may look.

While the mobile is undertaking a call, it is able to monitor the signal strength and BER from the signals surrounding base stations with the information provided from the measurement order. As the system uses TDMA, the mobile is able to undertake the measurements during the time when it is not transmitting or receiving. The results of these measurements (BER and RSSI – Received Signal Strength Indication) are returned to the base station as a Channel Quality message using the SACCH.

If the network decides that a handoff is required, then the mobile is allocated a new channel and moves to this. The actual channel allocation and base station is determined by the reports sent back to the network from the mobile.

Authentication

NA-TDMA was developed after the first-generation analogue systems were established. These systems were open to fraud, and there were many instances of this. To overcome this, preventive measures were introduced into the second-generation systems. Being digital, this was much easier to achieve.

The authentication processes were standardized within IS-41C, and this is independent of the air interface. Accordingly, this has also been also adopted for other schemes, including cdmaOne/IS-95. It consists of a several-step process, and is additionally undertaken at several different stages, including registration, call origination and call termination, and when the mobile moves to a new location.

The Authentication Centre (AuC) is at the core of the process. It is the database at the centre of the network where much of the private authentication data is stored. A 64-bit number known as the A key is stored in each mobile in a way that is cannot be known even by the user. It is also not transmitted over the air.

A number known as the Shared Secret Data (SSD) is used; this consists of a 128-bit number which itself consists of two 64-bit sections. These are known as SSD_A and SSD_B. The first of these, SSD_A, is used for authentication purposes, while SSD_B is used for voice encryption.

The SSD is generated using both the A key and the Electronics Serial Number (ESN) held within the mobile. This enables the SSD to be changed if the network feels that security may have been compromised.

By utilizing these techniques the authentication process and network security can be successfully maintained.

PDC

In Japan, the second-generation TDMA cellular system is very similar to NA-TDMA and is known as PDC. This stands for either Pacific Digital Cellular or Personal Digital Cellular. The standard was defined by RCR, which later became ARIB, the Japanese standards organization. The first service was launched in March 1993 by NTT DoCoMo, and by 2003 the system had over 61 million subscribers. The system is used nowhere else in the world, and is now being phased out with the advent of 3G services. Both UMTS and CDMA2000 are being rolled out.

PDC uses a 25-kHz carrier spacing with three time slots. The modulation used is pi/4-DQPSK, and this enables it to carry low bit-rate 11.2-kb/s and 5.6-kb/s (half-rate) vocoder data. The signalling is very similar to that employed on NA-TDMA. The bands used for the service are 800 MHz (downlink 810–888 MHz, uplink 893–958 MHz) and also 1.5 GHz (downlink 1477–1501 MHz, uplink 1429–1453 MHz). The services include voice (full- and half-rate), supplementary services (call waiting, voice mail, three-way calling, call forwarding and so on), data service (up to 9.6 kbps CSD) and packet switched wireless data (up to 28.8 kbps PDC-P).

cdmaOne/IS-95

While spread spectrum techniques had been used for many years in military circles, and then primarily for covert communications systems, it was not until 1989 that the idea of using direct sequence spread spectrum techniques in the form of a code division multiple access scheme was proposed to the Telecommunications Industry Association in the USA. As there was already another second-generation technology being developed, the industry did not want competing technology to be introduced. Accordingly, the first CDMA networks were not launched in the USA, from where the idea had emanated. Instead the first network was launched in Hong Kong in 1994, with the first network in the USA being launched in 1996.

The idea of using direct sequence spread spectrum techniques to provide an access technology was quite revolutionary. It was a major leap in technology over the original FDMA schemes used for the first-generation networks, and also the TDMA systems being rolled out as GSM, initially in Europe, and the NA-TDMA system being rolled out in the USA. It relied heavily on the much greater levels of digital signal processing power that were coming and as such formed a significant development in the approach used for mobile communications.

As described in Chapter 3, CDMA relies on the use of different orthogonal spreading codes to differentiate between the different mobiles accessing the system. Although the signals all occupy the same frequency channel, the system is able to decipher each signal by applying the correct code to correlate the incoming signal and decipher it.

Standards

The first CDMA system was defined under IS-95. The first revision, termed IS-95A, was released in 1995, and specified the air interface for a cellular, 800-MHz frequency band. ANSI J-STD-008 specified the PCS version (i.e. the air interface for 1900 MHz). Later, the IS-95 standard was revised to give IS-95B. This merged the IS-95A, ANSI J-STD-008 and others. Additionally, it specified 'high-speed' data operation with a maximum bit rate of 115.2 kbps. Although the system was defined by a number of standards, it was marketed under the brand name cdmaOne. It is this name that will be seen in any sales information.

Spreading codes

As a CDMA system is based on direct sequence spread spectrum techniques, the core of the system revolves around the use of the orthogonal chip or spreading codes. The codes used in cdmaOne are called Walsh codes, and the system combines these with two pseudo-random noise (PN) codes for each communication channel. By using a number of different codes, it is possible to create different channels. These can either be used for control applications or for carrying the data payload. The different channels can then be separated by the receiver, as the receiver is only able to correlate those for which it has been provided with the required Walsh codes. All other signals using other Walsh codes just appear as noise.

The Walsh codes used for cdmaOne have a fixed length of 64 bits, and this allows for a maximum of sixty-four individual physical channels to be carried by the system. The maximum number of usable channels is limited by the level of interference from outside, possibly adjacent cells, and the data throughput on each channel. This means that as interference from neighbouring cells rises so do the interference levels, and the capacity of a given cell becomes less. This may mean that it is not possible to utilize all the channels, and less than the maximum sixty-four may be used.

A further type of code, known as a PN code, is used in assembling the signal. Called pseudo-random noise codes, they consist of a series of bits that have random properties. Three PN codes are used within cdmaOne, two short codes and one long one.

These codes are used because they have some particularly useful properties. If one code is time-shifted, it is found that the two copies of the same code become uncorrelated. This means that the same code can be used twice, but time-shifted to provide two uncorrelated sequences. In fact, when used in the network the codes are synchronized to triggers that are synchronized to GPS timing signals.

In order to utilize the PN codes, an approach using a mask is adopted. Codes transmitted at a different time offset are uncorrelated, and therefore by adopting a known given set of offsets it is possible to create a set of 'codes' that can be used. The short codes are 32 768 bits long and offset by 64 bits. This gives a total of 512 offsets that can be used. As the short code is transmitted at the CDMA data rate of 1.2288 Mbps, this equates to a full cycle time of 26.667 ms.

As will be seen in later explanations, a process known as 'decimation' is used when generating the channels. This is a process where only part of the data acquired is actually used. If only 1 out of 5 bits is used, then the sequence is decimated by a factor of 5:1.

Radio signal construction

CDMA transmissions are very different to those used for the previous FDMA and TDMA systems. Here, the system occupied a relatively narrow bandwidth. For a CDMA system, a much wider bandwidth is used. In the case of IS-95/cdmaOne, a channel bandwidth of 1.25 MHz is used. The actual bandwidth is determined by the chip rate, which is 1.228 Mcps. Additionally, in another departure from previous practice, adjacent cells are able to use the same frequency, the different codes used by the adjacent cells enabling them to work alongside each other.

The characteristics of the transmission between the forward and reverse links are slightly different. Much of this arises from the fact that the two links are on different frequencies, and the path lengths are different in terms of the number of wavelengths travelled. In the forward link, a form of modulation known as Quadrature Phase Shift Keying (QPSK) is used. This allows 2 bits per symbol to be transmitted, and is therefore an efficient method of transmitting the data. In the reverse channel, a form of modulation known as Offset Quadrature Phase Shift Keying (OQPSK) is used. This is a form of quadrature phase shift keying where the quadrature (Q) component of the signal is offset from the in-phase (I) component by half a symbol period. As a result the amplitude of the resulting waveform does not cross zero, and this reduces the peak-to-average ratio. This has advantages for power amplifiers because the reduction enables more efficient signal amplifiers to be made in the mobile, where battery consumption is a key parameter.

Channels

The channels in cdmaOne can be split into those used for the forward link and those used for the reverse link. The channels between these two links differ slightly, and will be treated separately. The reason that the uplink and downlink are treated differently is that Walsh codes need to be synchronized if they are to remain orthogonal. As the signals transmitted from the mobile stations travel over different distances because of the variety of locations of the mobiles, they will all arrive at slightly different times and hence will not be synchronized with one another. The addition of the PN code overcomes this problem, even though Walsh codes have been used for part of the spreading process in the mobile.

Forward link code channels

There are sixty-four code channels that can be accommodated on each CDMA RF channel. Of these there are three overhead channel types: the pilot channel, the paging channels (of which there may be up to seven) and the Sync channel. Beyond these channels, the remaining codes can be traffic channels. Their arrangement is shown in Figure 8.1.

The data from the individual channels are multiplied with the Walsh codes to provide the individual channels. The product is then further multiplied with the short PN codes explained above (Figure 8.2). This has the effect of acting as an identifier for the sector/cell in question. In this way, the mobile can determine from which cell the signal is coming.

Pilot channel

The Pilot Channel (PC) acts as a timing beacon for the system, and as a result it carries no data. In fact the data is a stream of zeros, and these are spread using Walsh code 0, which is also a

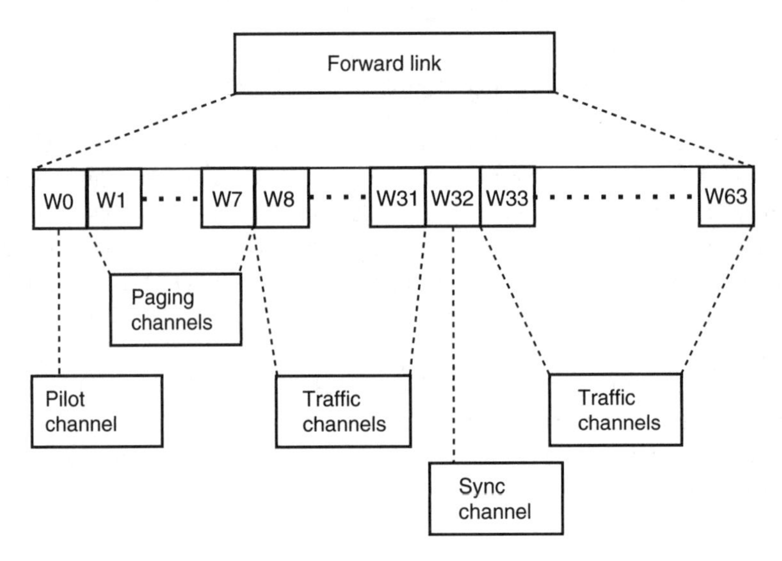

Figure 8.1 Forward link channels.

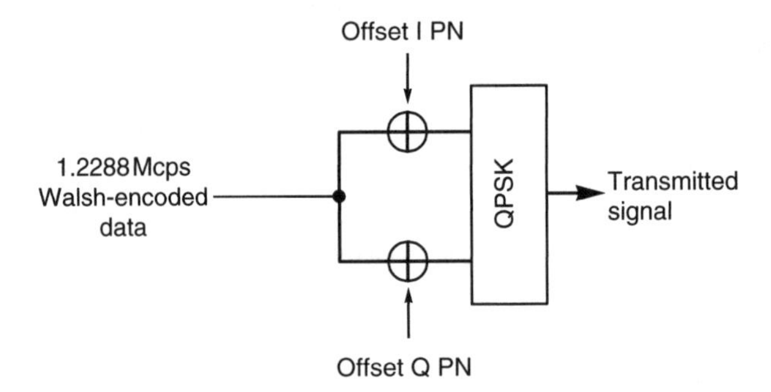

Figure 8.2 Generation of the transmitted signal showing how the Walsh-encoded channels are further spread with PN codes. The offset on the PN codes identifies the base station/sector.

sequence of zeros. The resulting sequence, again a string of zeros, is spread by a pair of quadrature PN sequences (Figure 8.3). This means that the pilot channel is effectively the PN sequence with its associated offset. By doing this, the channel is used by the mobiles as a phase reference as well as a method of identifying the cell. Both the Walsh code and the PN sequence run at a rate of 1.228 Mcps.

Once the signal has been spread by the PN sequence, baseband low-pass filters are used to shape the data stream and in this way the baseband spectrum of the signal is controlled, enabling the overall signal bandwidth to have a sharp roll-off at the band edges.

The pilot channel is transmitted continuously by each sector of a base station. It provides the mobile with timing and phase reference. A measurement of the signal-to-noise ratio of the pilot channel also gives the mobile an indication of which is the strongest serving sector.

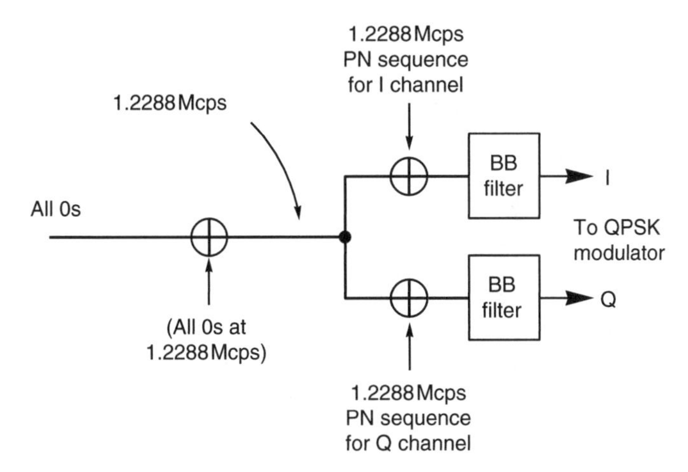

Figure 8.3 Pilot channel generation.

Paging channels

As the name indicates, the Paging Channels (PCH) carry information, primarily to page or call mobiles. This includes system parameters, voice pages, short message service and any other broadcast messages to users in the cell. The channels can occupy Walsh codes 1 through to 7, dependent upon the system requirements. Walsh code 1, i.e. channel 1, is always used as a minimum.

The paging channel carries data at either 4.8 or 9.6 kbps. A field in the Sync channel informs the mobile of the data rate of the paging channel, and after the mobile has gained synchronization with the base station it monitors the paging channel for messages. Although there can be up to seven paging channels per sector, each mobile only monitors one paging channel, with the channel and time interval being determined on registering with the network.

As with other channels, there are several stages in producing the final signal. First, the baseband information is error protected. After this, the data are repeated if at a rate of 4.8 kbps; otherwise they are left alone. Following this, the data are interleaved and then scrambled by the decimated long PN sequence, and finally spread by the Walsh code for the particular channel assignment (Figure 8.4). In this process, the long PN code is itself masked with a code which is specific to the channel being used. In this way, the long PN code for paging channel 1 (using Walsh code 1) is different from that for paging channel 4 (using Walsh code 4).

To enable idle mobiles to be able to switch their receivers off for periods of time, the paging channel is split into slots and frames. In this way the data can be partitioned. The channel is divided into slots that are 80 ms in length. These are grouped with 2048 slots, being called the 'maximum slot cycle'.

Moving in the other direction, each slot is divided into four paging channel frames, and each paging channel frame is further divided into two paging channel half-frames. Within each of the half-frames, the first bit is known as the synchronized capsule indicator (SCI) bit.

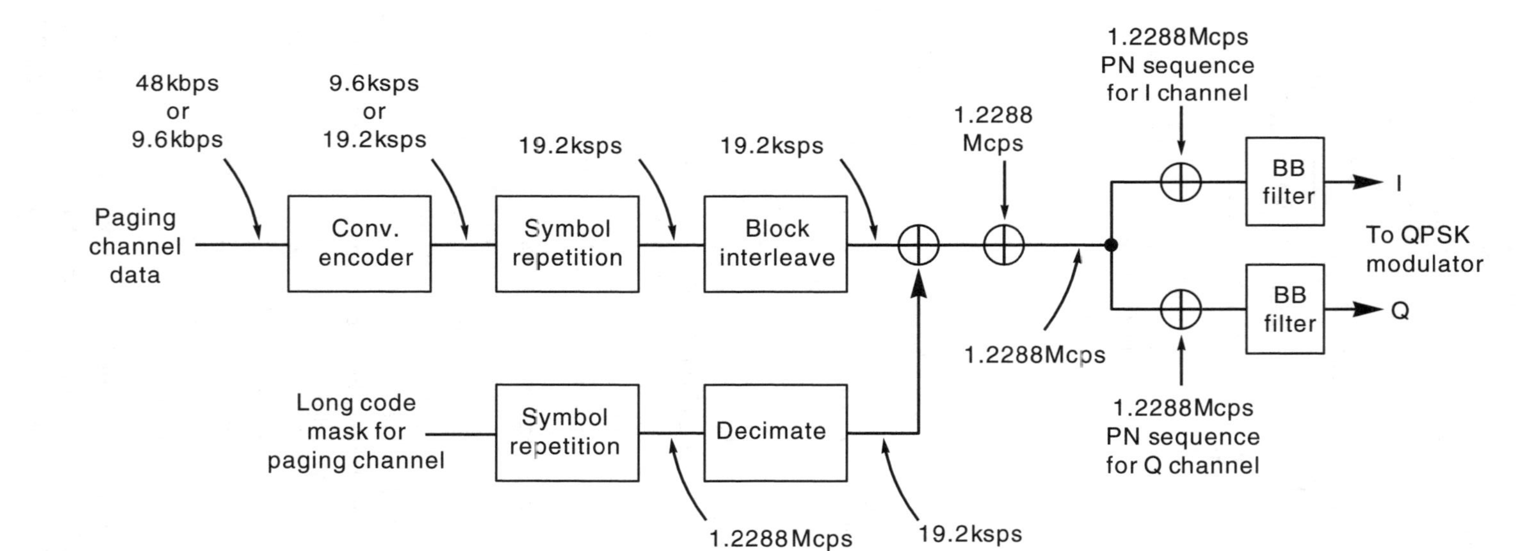

Figure 8.4 Paging channel generation.

The messages sent on the paging channel may vary in length. Accordingly, they may occupy more than one paging channel half-frame and even end in the middle of a half-frame. The message may be carried by one of two methods: synchronized paging channel message capsules or unsynchronized paging channel message capsules.

When a message may end in the middle of a half-frame, then the base station needs to ensure that the message adopts specific formats so that the mobile can decode it properly. If there are less than 8 bits between the end of that message and the SCI bit of the next half-frame, then the base station includes 'padding' bits to fill the remaining space.

If the message ends in the middle of a paging channel half-frame and there are more than 8 bits between the end of that message and the SCI bit of the next half-frame, it is handled differently. Under these circumstances, the base station may transmit an unsynchronized message capsule immediately after the message. In this case, no padding bits are added and the SCI bit flags the start of a brand new message capsule in the current half-frame. In other words, a new message capsule starts immediately following the SCI bit. For all other cases, the SCI bit is set to 0.

Figure 8.5 illustrates the paging channel structure.

Synchronization channel

The Synchronization Channel (SC) always uses Walsh code 32, and it is primarily used by the mobile to acquire the timing reference. The timing is very important, as the base stations are kept in tight synchronism with each other by a signal derived from the Global Positioning Service (GPS). Each base station has a fixed timing offset from the GPS signal to minimize interference between cells, and the synchronization channel provides the mobile with the required timing offset information to access the cell.

The sync channel incorporates an 80-ms superframe that is divided into three 26.667-ms frames which correspond to the same length as the short PN sequences. Accordingly, they align with the timing on the pilot channel. In terms of power, the SC is allocated the least power of the overhead channels in the overall CDMA transmission.

The data transmitted on this channel include the system time, pilot PN of the base station, long code state, system ID and network ID.

Figures 8.6 and 8.7 illustrate sync channel coding and the sync channel framework, respectively.

Forward traffic channel

The primary purpose of the Forward Traffic Channel (FTC) is to transmit voice. However, voice does not require a constant bit rate when vocoded, and IS-95 allows the rate of the frames to change dynamically (every 20 ms). When the rate is decreased, it reduces the level of interference to other users. The original vocoder specification used a set of rates based on divisions of 9.6 kbps, and this is reflected in IS-95A. Later the vocoder was improved to give better voice quality, and in

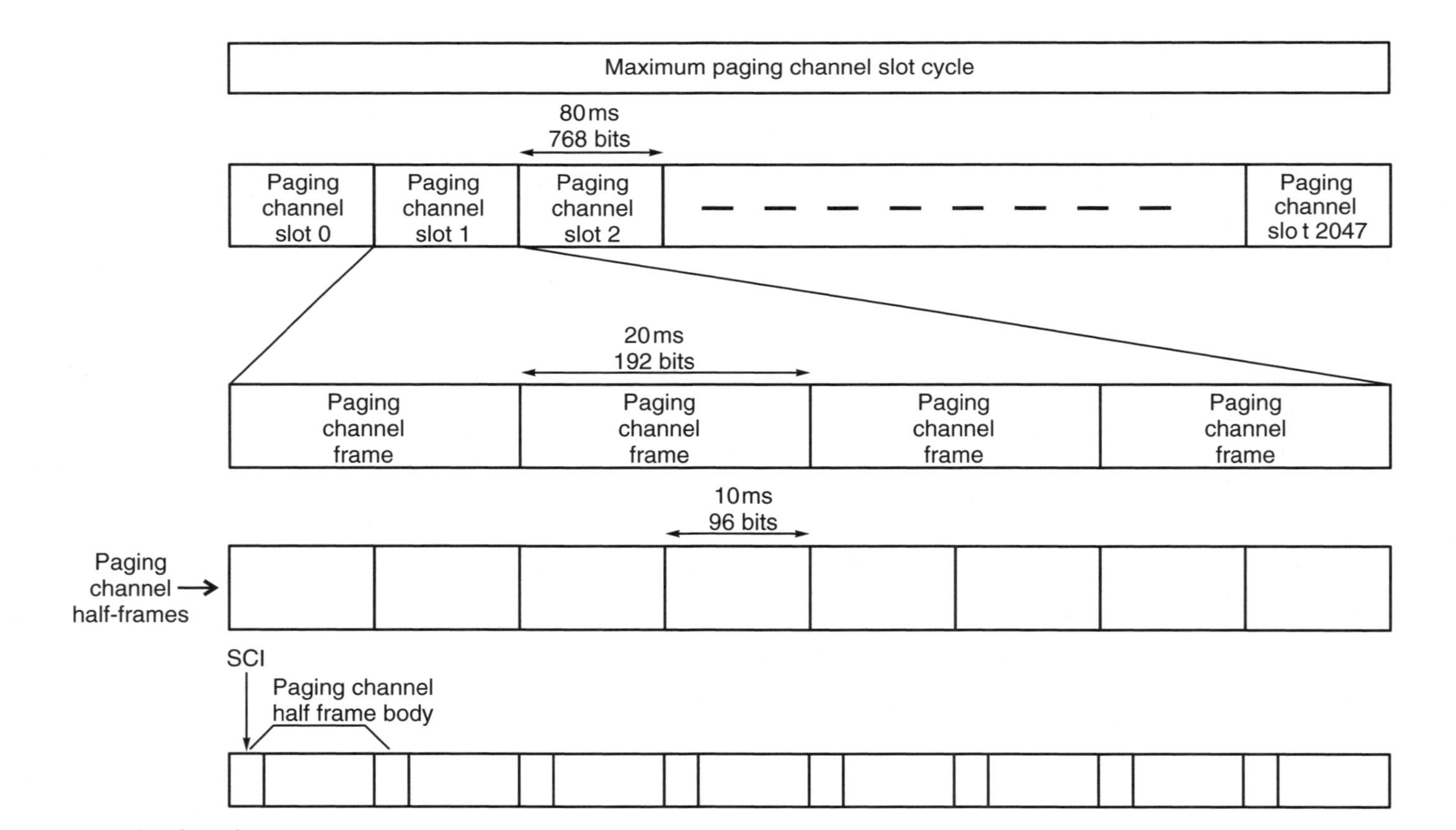

Figure 8.5 Paging channel structure.

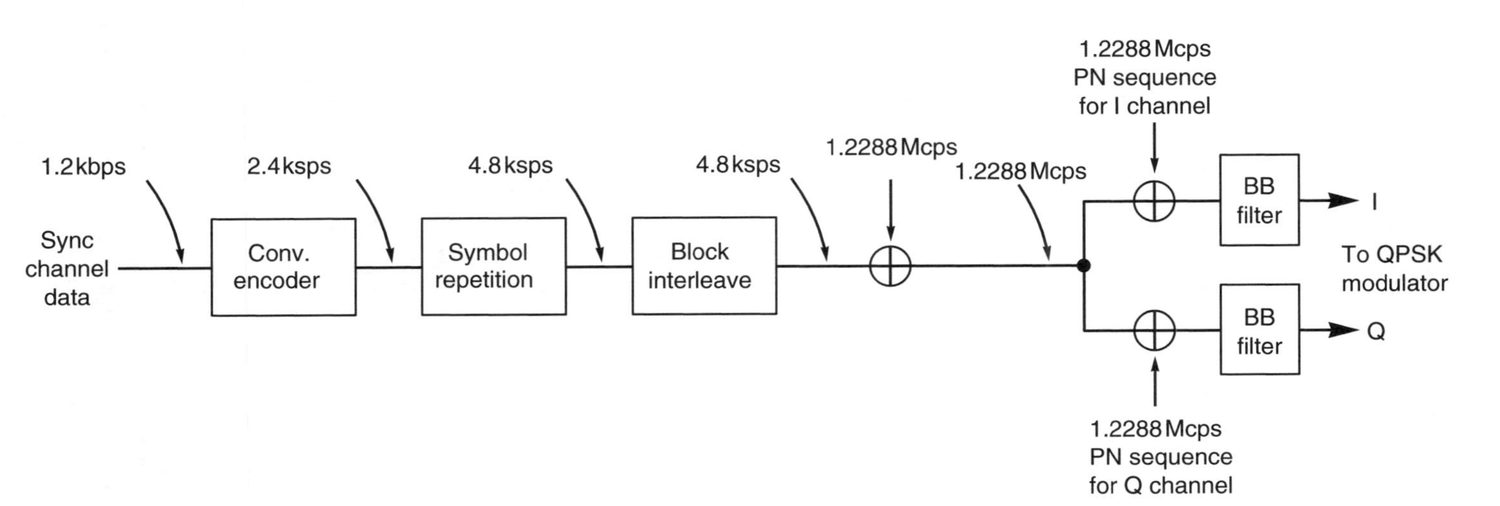

Figure 8.6 Sync channel coding.

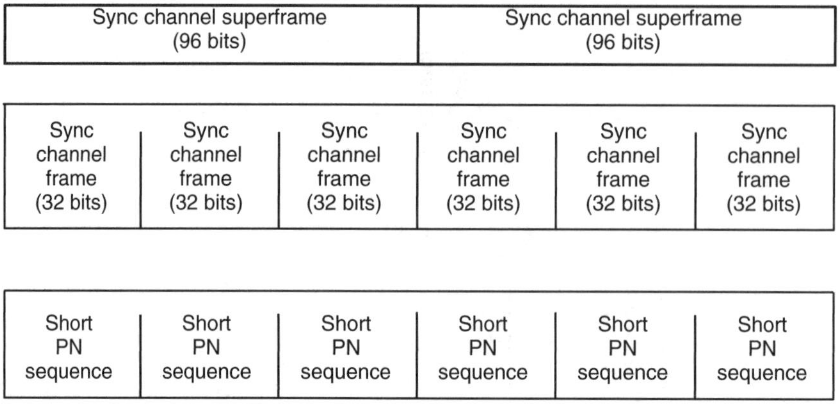

Sync channel superframe (96 bits)	Sync channel superframe (96 bits)

Sync channel frame (32 bits)	Sync channel frame (32 bits)	Sync channel frame (32 bits)	Sync channel frame (32 bits)	Sync channel frame (32 bits)	Sync channel frame (32 bits)

Short PN sequence	Short PN sequence	Short PN sequence	Short PN sequence	Short PN sequence	Short PN sequence

Figure 8.7 Sync channel frame structure.

IS-95B a vocoder was introduced with a rate set based on 14.4 kbps. The 9.6-kbps rate set was termed RS1, and the second, based on 14.4 kbps, was termed RS2. However, data are always carried at full rate.

The forward traffic channel is used to transmit user voice and also data. Signalling messages are also sent over it. The basic structure of the forward traffic channel is similar to that of the paging channel, and it can be seen in Figure 8.8 for RS1. Here, the vocoder varies its output data rate as a result of the varying levels of speech activity from 9.6 kbps down to 1.2 kbps. Obviously, the lowest is used during periods of quietness – i.e. pauses in the speech.

The baseband data from the vocoder is encoded with error protection, and for RS1 a rate 1/2 convolutional encoder is used, which effectively doubles the bit rate. Next, the data undergo symbol repetition. This has the effect of reducing the level of interference on the radio channel, and it is introduced when a lower data rate is produced by the vocoder, as detailed below.

For a vocoder data rate of 9.6 kbps, the code symbol rate is 19.2 ksps at the output of the convolutional encoder and no repetition is introduced. For a data rate of 4.8 kbps, the code symbol rate is 9.6 ksps and each symbol is repeated once, to produce a final output symbol rate of 19.2 ksps. For the next rate down, 2.4 kbps, the code symbol rate is 4.8 ksps and each symbol is repeated three times, again giving an output rate of 19.2 ksps. Finally, for a data rate of 1.2 kbps, the code symbol rate is 2.4 ksps and under these circumstances each symbol is repeated seven times.

After symbol repetition, the data are interleaved to combat burst errors. This process involves taking the data, splitting them into blocks and then grouping sections of data from different blocks together. In this way, if a block of data is lost then it is possible to reconstruct the data more easily. The interleaved data are then scrambled, using the long PN sequence.

The long PN generator outputs a long PN sequence at 1.2288 Mcps and it uses a mask from the mobile Electronic Serial Number (ESN) of the mobile, providing an identity for the signal.

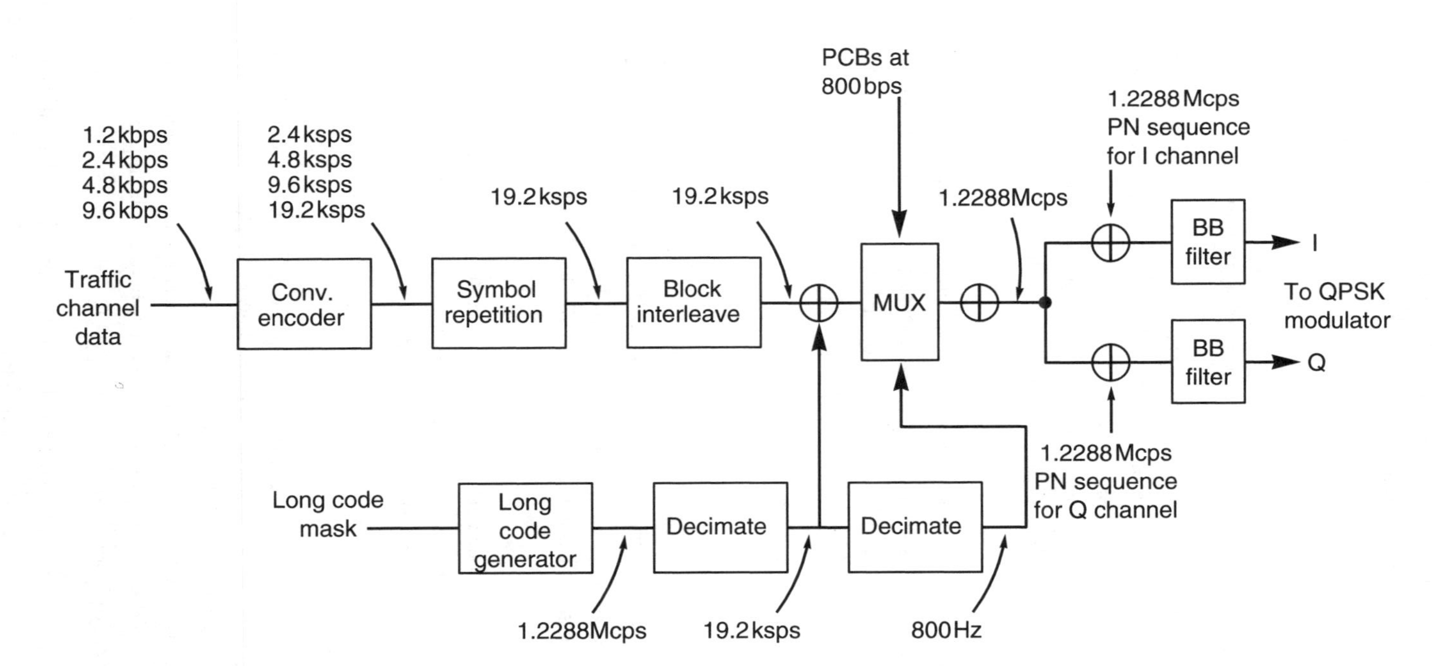

Figure 8.8 Forward channel generation.

To enable the PN sequence rate to be reduced to 19.2 ksps, it is decimated by a ratio of 64:1 and multiplied with the interleaved data. Additionally, power control bits at a rate of 800 bps are multiplexed with the scrambled data.

The next stage in the process is for the data stream in the forward traffic channel to be spread using a Walsh code. Each forward traffic channel is spread using a different Walsh code, and in this way it can be identified and separated in the receiver. This may be W8 to 31, or W33 to 63. W2 to W7 may also be used if they are not being used as paging channels.

With the Walsh code chip rate of 1.2288 Mcps; each symbol is spread by a factor of sixty-four and the result is a spread data stream at a rate of 1.2288 Mcps. The resulting data stream is further spread by a short PN code, one of which is assigned for a particular sector. The short PN sequence provides a second layer of isolation, and distinguishes among the different transmitting sectors. By doing this, all the sixty-four available Walsh codes can be re-used in every sector.

The forward traffic channel structure is similar for RS2, although there are several differences. One is that the vocoder codes speech at higher rates, and provides better voice quality. It supports four variable rates, of 14.4, 7.2, 3.6 and 1.8 kbps. In order to maintain the output of the block interleaver at 19.2 ksps, the rate of the convolutional encoder is increased to $R=3/4$.

Modulation

Once all the individual channels have been assembled, they need to be summed and modulated onto the radio frequency carrier. To achieve this the gain of each logical channel is adjusted using a separate gain control for each channel, and in this way the relative powers for each channel can be set. This is required because the levels for the traffic channels are constantly changing.

Once the channel gains have been set, the signals are added coherently to form the spread spectrum signal. The I and Q components are then each modulated onto a carrier and summed to give the final transmitted QPSK signal (Figure 8.9).

Reverse channels

For the reverse link, the channels are set up in a slightly different manner. There are only two basic channels, namely the Access Channel and the Reverse Traffic Channel. The way in which they are structured and assembled is also different in view of the fact that they are generated in the mobile rather than the base station. One of the differences already mentioned regards the type of modulation used. Rather than QPSK, OQPSK is used, where a half-chip delay is introduced onto the Q channel of the modulation.

Orthogonal modulation schemes are again used on the reverse link. The different mobiles are identified by a mask based on the Equipment Serial Number (ESN) on the long PN code. This is used to give the final spreading of the data to 1.228 Mcps.

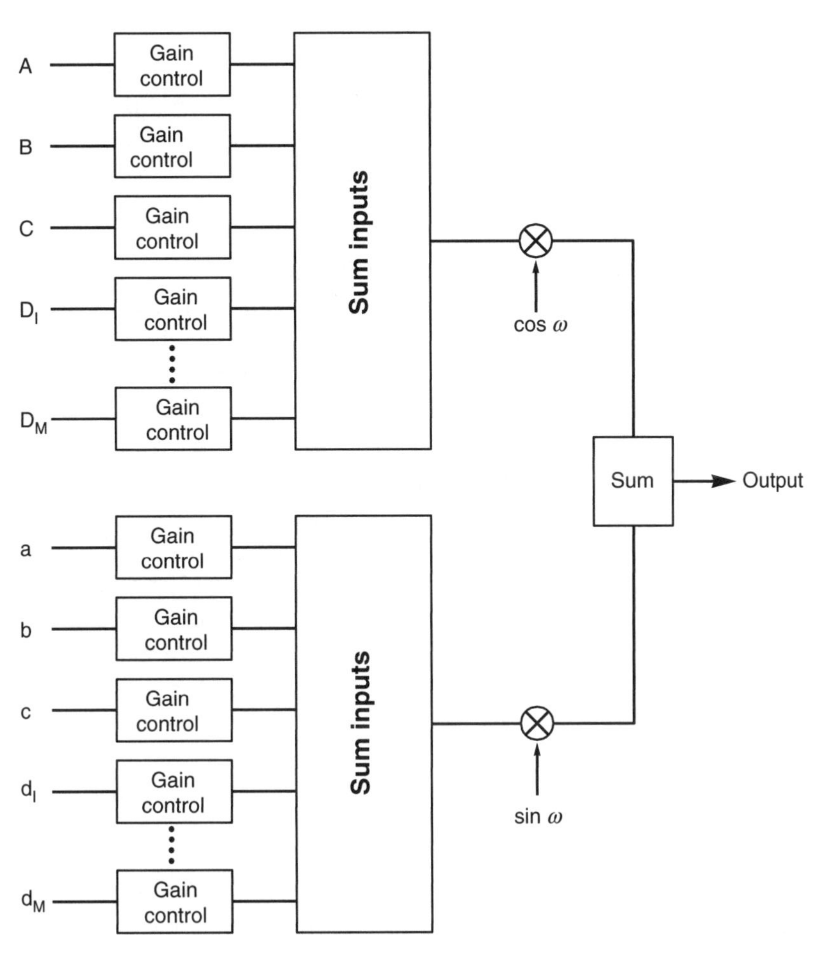

Figure 8.9 Modulator used for the forward link.

The mobile transmits for a portion of the time, and this not only increases the battery life but also serves to keep the overall level of interference to a minimum to ensure that the cell capacity is kept as high as possible.

Access channel

The Access Channel (AC) is used by the mobile to communicate with the base station when no traffic channel has been set up. It therefore uses the channel for call origination requests, and also to carry the responses to paging commands sent by the network to the mobile.

There can be up to 32 access channels on the reverse link for each paging channel on the downlink. Each AC uses the PN, but time-shifted to enable it to be uniquely identified. Data are sent at 4800 bps in a 20-ms timeframe, so that each frame contains 96 bits.

Modulation

To generate the signal, the baseband information is first error protected using a convolutional encoder. Symbols are then repeated once to give a code symbol rate of 28.8 ksps. The data are interleaved and then coded using a 64-ary orthogonal modulator.

The set of sixty-four Walsh functions is used, but here the Walsh functions are used to modulate groups of six symbols. The reason for orthogonal modulation of the symbols arises from the fact that the mobiles are not synchronized in the same way as are the transmissions from the base station. This means that the signals arriving at the base station will not be coherent, because the receiver in the base station will receive signals with different timings due to small timing errors in the mobile and also the different transmission times resulting from the different path lengths from the mobiles to the base stations. When the transmission of a mobile is not coherent, the receiver (at the base station) still has to detect each symbol correctly. Making a decision as to whether or not a symbol is +1 or −1 is difficult when judged over the period of one symbol. To overcome this, if a group of six symbols is represented by a unique Walsh code, then the base station can easily detect six symbols at a time by deciding which Walsh function is sent during that period. Unlike the forward link, where the Walsh codes are used to distinguish between different channels, on the reverse link they are used to distinguish between different groups of six symbols. Thus the receiver can easily identify which Walsh function is being transmitted by correlating the received sequence with the set of sixty-four known Walsh functions. In other words, the six-symbol pattern dictates which Walsh code is used. If a pattern of +1, −1, −1, +1, +1, +1 is to be sent, this corresponds to a binary value of 100111, which has a decimal value of 39. Accordingly, Walsh code 37 is used.

The orthogonally modulated data are then spread by a long PN sequence running at a rate of 1.2288 Mcps, which is used to distinguish the access channel from the channels in the reverse link. The data are further scrambled in the I and the Q paths by the short PN sequences running at 1.2288 Mcps. Data in the Q leg are delayed by half a chip to give OQPSK modulation (Figure 8.10).

Information is transmitted in access channel slots and access channel frames. Each frame contains 96 bits and lasts 20 ms, giving a baseband data rate of 4.8 kbps. The data comprise 88 information bits and 8 encoder tail bits. There are two types of messages sent on the access channel: a response message (in response to a base station message) or a request message (sent by the mobile station).

Reverse traffic channel

The Reverse Traffic Channel (RTC) carries variable rate data consisting of voice, data, control information, etc. The frames contained within the channel are variable-rate frames at 4800, 2400 or 1200 bps.

The structure of the reverse traffic channel is similar to that of the access channel, but additionally it includes a data burst randomizer into which the orthogonally modulated data are fed. The data burst randomizer is the technique by which the voice activity factor is accounted for on the reverse

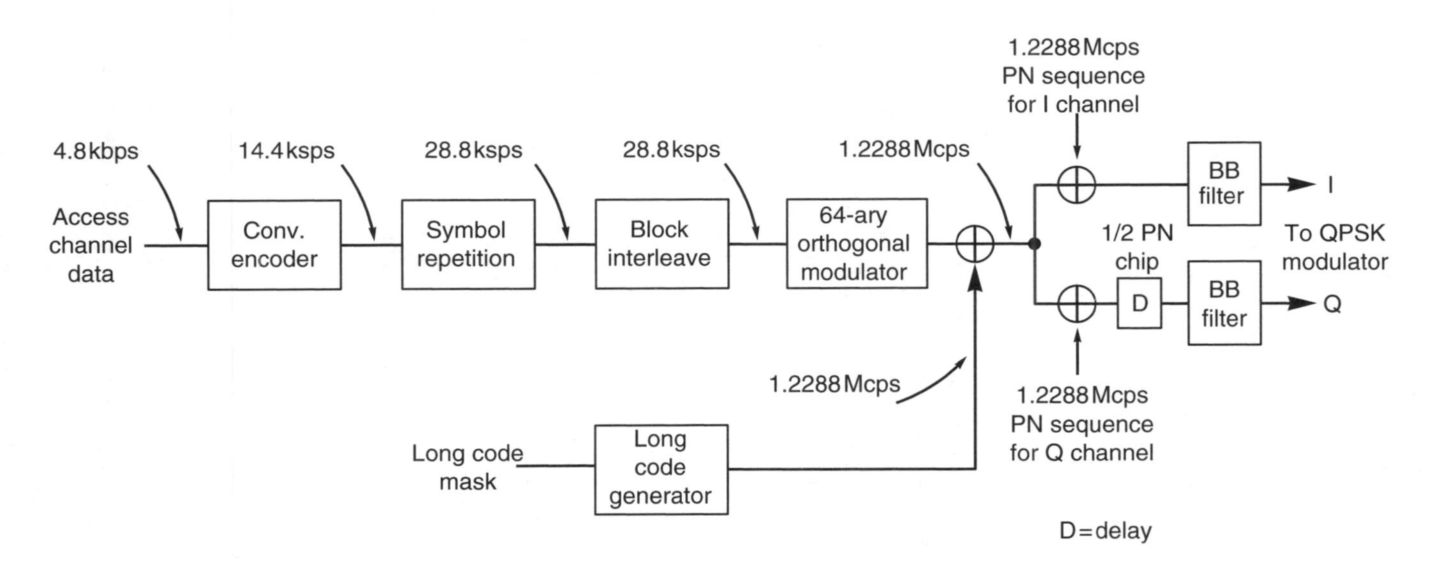

Figure 8.10 Reverse access channel generation.

link. It is not possible to use the same techniques used on the reverse link that are used on the forward link because they affect the channel power. This would upset the power control adjustments that need to be made to ensure that all the mobiles are received as close to the same strength as possible. Accordingly, a method that does not upset the measurements taken in the base station needs to be adopted.

For the reverse link, the technique used is pseudo-randomly to switch off the transmitter during redundant symbols produced by symbol repetition. This is accomplished by the data burst randomizer. It first creates a mask that randomly masks out redundant data. This is partially governed by the vocoder rate. If the vocoder is operating at 9.6 kbps then no data are masked, but if, for example, the vocoder is operating at 1.2 kbps then the symbols are repeated seven times, and the data burst randomizer masks out seven of eight groups of symbols.

The traffic channel frames are divided into sixteen Power Control Groups (PCG), each 1.25 ms in length, and accordingly the data burst randomizer pseudo-randomly masks out individual power control groups to enable the power control to be maintained correctly. When the vocoder operates at 9.6 kbps, no PCGs are masked out; when the vocoder operates at 4.8 kbps, an average of eight PCGs are masked out in a frame. While the masking pattern is mainly dependent upon the vocoder rate it is also dependent upon the long PN sequence used to spread the previous frame, and this is used in determining the masking pattern.

Like the forward traffic channel, the reverse traffic channel frames are 20 ms long. At full rate it contains 192 bits, i.e. a data rate of 9.6 kbps, while at half rate it contains 96 bits, giving a data rate of 4.8 kbps, etc. The full-rate and half-rate frames contain frame quality indicator Cyclic Redundancy Check (CRC) bits, and all frames contain encoder tail bits.

The traffic channel is used to carry voice and data traffic as well as signalling. In fact, the system can multiplex primary or voice traffic with the secondary or signalling traffic. There are two techniques that are used, although only on full-rate frames. The first is known as 'blank and burst'. Here, the entire traffic channel frame is used to send only secondary data, while the entire traffic channel frame is used to send only signalling data. The secondary or signalling data effectively blank out the primary data. The second technique is known as 'dim and burst'. With this the traffic channel is used to send both primary and secondary data, although the primary traffic is only sent at half rate – i.e. 80 bits per frame – or sometimes more.

Power control

In order for the receiver to receive all the incoming signals at approximately the same strength, power control is of great importance. Accordingly, IS-95 provides three different power control mechanisms. Power control is particularly important in the reverse link to the base station, where it is necessary to keep all the mobiles at approximately the same strength so that the stronger mobiles do not mask the weaker ones in an effect known as the near–far effect. Accordingly, both open-loop and fast closed-loop power control are employed. In the downlink, a relatively slow power control loop controls the transmission power.

Open-loop power control

Open-loop power control provides a broad level of power control. Accordingly, it is used to adjust the initial access channel transmission power of the mobile station and to compensate for large, abrupt variations in signal strength. To use it, the mobile station determines an estimate of the path loss between the base station and mobile by measuring the received signal strength at the mobile. This measurement is then used to determine the power level the mobile will use when accessing the system.

Closed-loop power control

The forward and reverse links have a frequency separation of 45 MHz, and as a result the fading and many of the path loss variations do not correlate well. Even though the average power is approximately the same, the short-term power resulting from small changes in position, etc., is different. This means that the open-loop power control mechanism cannot accurately set the required power level to give the correct signal at the base station, and hence the closed-loop power mechanism is required. To achieve this, the base station measures the received signal-to-interference ratio (SIR) over a 1.25-ms period, which is equivalent to twenty-four modulation symbols. It then compares this to a target SIR, and decides whether the mobile station transmission power needs to be increased or decreased. The power control bits are transmitted to the mobile on the downlink traffic channel every 1.25 ms, giving a transmission rate of 800 Hz, and this is achieved by puncturing the data symbols. The placement of a power control bit is randomized within the 1.25-ms power control group. The transmission occurs in the second power control group following the corresponding uplink traffic channel power control group in which the SIR was estimated. The power control commands are transmitted uncoded, and this is why they are transmitted at full-rate power, regardless of the frame rate. However, any errors are made less critical by the fact that the power is continually monitored and altered accordingly, up or down. Thus errors in the power correction data are soon compensated.

On receipt of the power control commands, the mobile extracts the power control bit commands and adjusts its transmission power accordingly. The adjustment step depends upon the system, and may be 0.25, 0.5 or 1.0 dB. The dynamic range for the closed-loop power control is ±24 dB. The composite dynamic range for open and closed loop power control is ±32 dB for mobile stations operating in the North American cellular band around 850 MHz, and ±40 dB for mobile stations operating in the PCS band around 1900 MHz.

Downlink slow power control

Although the power control in the downlink is less critical than in the uplink, it is nevertheless important. It serves to improve the performance of mobile stations at a cell edge where the signal is weak and the interfering base station signals are strong. The base station periodically reduces the transmitted power to the mobile station. For this control, the mobile measures the Frame Error Ratio (FER) for the received signal. When this exceeds a predefined limit, typically 1 per cent, the mobile station requests additional power from the base station. This adjustment is not as swift as that used for the reverse link, and occurs every 15–20 ms. The dynamic range is also limited, with an adjustment range of only ±6 dB.

Handoff

One of the advantages of CDMA systems is associated with handoff. As a result of the fact that the same frequency is used for adjacent cells, mobiles are able to receive signals from more than one base station at the same time. Normally they would reject all but the signal from the wanted base station as part of the decoding process. However, the ability to receive more than one signal is advantageous when handing off from one cell to the next. During the handoff process, the two adjacent cells both send the same data to the mobile. The mobile listens to both signals, and when the handoff is complete the mobile returns to listening to just the one base station. This approach significantly reduces the possibility of a call being dropped during handoff, and it also ensures optimum voice quality when a mobile is on the edge of a cell. This form of handoff is known as a 'soft' handoff.

Whilst soft handoff appears to be a significant advantage, there is a downside to the process. The first is that it reduces the system capacity. The reason for this is simply that it requires two channels or codes, one for each base station. It is estimated that this can reduce system capacity by as much as 40 per cent, but this is dependent upon many factors, including the duration of the handoff, the size of the cells and the level of overlap between the cells. There are disadvantages for the mobile as well in that it requires two receivers or demodulators to be able to monitor the two cells. Whilst this adds complexity to the mobile, it can be overcome by integrating the additional functions into the main IC, and therefore the cost is relatively small. Despite these disadvantages, the advantages of a much smoother and more reliable handoff are worth the additional costs.

There is another form of handoff called a 'softer' handoff. This occurs when a mobile moves between sectors on the same base station. As elements of the processing are shared, this enables a 'softer' handoff to be accomplished even more easily than a soft handoff.

It may also be necessary for a mobile to undergo a 'hard' handoff. Under these circumstances, the mobile is instructed to change channels. This is considerably more involved than the soft handoff, because it is necessary for the mobile physically to break the connection with one base station before connecting to the next. Hard handoff is also required when handing off from IS-95 to an analogue network.

Discontinuous reception

In order to conserve battery capacity, the mobile is able to enter a sleep mode when it is idle and only listens for paging data periodically. Under these conditions the mobile shuts down any unnecessary areas of the circuitry, the receiver being one, and wakes them up only as required. In this way, the current drain on the battery can be minimized.

The mobile will undertake tasks such as monitoring the paging channel. There is a cycle set up to do this, and it is in multiples of 1.28 seconds with a maximum period of just over 168 seconds. The parameters for the sleep mode are set up when the mobile registers with the base station. In this way, the mobile does not miss any messages being sent to it.

Call processing

Call processing is the complete process that enables calls to be successfully routed through to a mobile to enable users to communicate with another user at some other location, whether he or she is on another mobile phone or connected to a land line.

In IS-95, there are four states that the mobile passes through from power to accepting a call. Although the process may appear somewhat simple, it needs to capture all the errors that may occur in real life, and this complicates the picture considerably. However, to gain an overview the main steps will be addressed in turn, and the host of possible error conditions will not be addressed here.

The four states are termed MS initialization, MS idle state, system access state, and finally, MS traffic state (Figure 8.11).

The first state that the mobile enters after it is powered on is known as initialization. When entering this state the mobile performs a cell search, listening for the primary paging channel on Walsh code 1. From the information it gains, it is able to determine which of the cells is the strongest and the most suitable to access. The mobile also acquires the pilot channel and then, using the sync channel, it gains synchronization with the base station, synchronizing its long PN code and system timing with that of the base station. It also registers with the network, and is allocated a time period or slot for monitoring the paging channel. The length of this slot is determined between the base station and the mobile.

Having achieved initialization, the next phase is for the mobile to be able to enter the idle state – the name being given because to the user the mobile appears to be idle. In reality, it is undertaking a large number of processes. The mobile periodically listens to the paging channel, as described above, but to reduce the drain on the battery the mobile powers much of the circuitry down, monitoring the paging channel only when pages are likely to appear. The mobile also undertakes various acknowledgement procedures, and re-registration if required. This may be necessary in a number of circumstances, such as after a certain time has elapsed, or when the mobile moves to a new cell.

The mobile enters what is known as the access state when activity commences. This may be to receive an incoming call or to initiate an outgoing call. During this phase the network determines which channels are available, and then the mobile is allocated a traffic channel on receiving what is termed a 'traffic assignment message'.

Once the resource for the call has been committed, it enters the traffic state. During this state or phase, traffic is exchanged between the base station and the mobile.

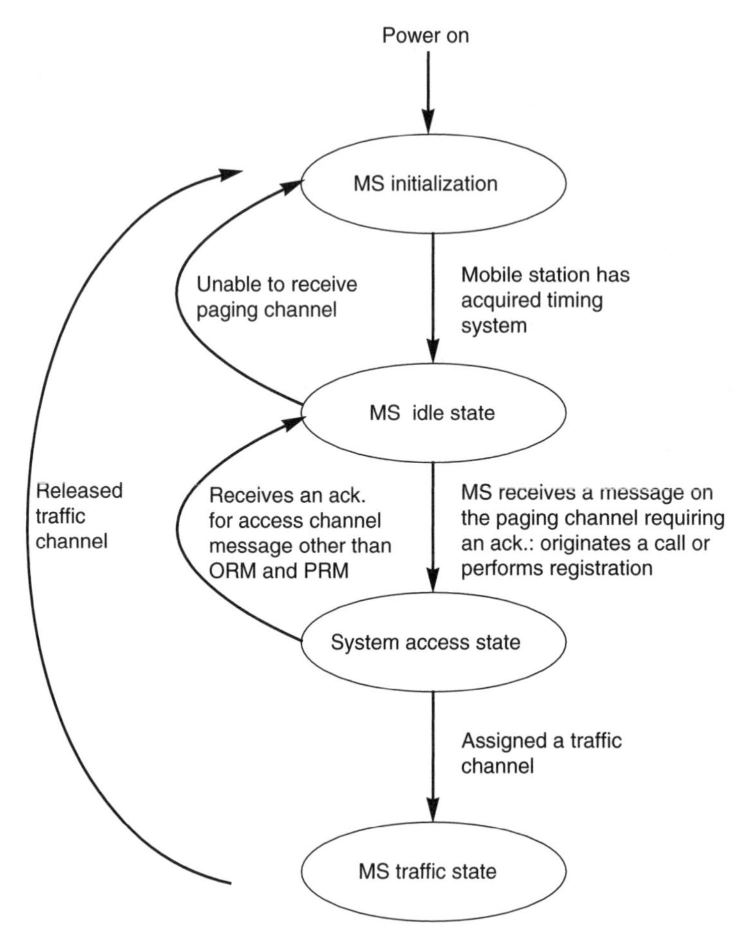

Figure 8.11 Basic four states of call processing.

Once the call is terminated, the mobile returns to its idle state, monitoring the paging channel for any requests that may appear. It may also send an access request if it is required to initiate a call.

Vocoders

There are three main vocoders used in cdmaOne. The first to be launched, known as the 8-K vocoder, operated at 8 kbps. This did not give a particularly good voice quality in view of the limited number of bits available, but it did make good use of the available capacity, thereby allowing operators to carry more calls. In view of the voice quality, another vocoder, the 13-K vocoder operating at 13 kbps, was introduced. This provided far better voice quality, but was not nearly as efficient in terms of the usage of the capacity. To overcome this problem, a third vocoder, called the Enhanced Variable Rate Codec (EVRC), was introduced. This is an 8-kbps vocoder that varies the compression of the voice signal into one of four data rates based on the rate of the user's speech activity. The four rates are full, 1/2, 1/4 and 1/8. The vocoder uses its full rate when a person is talking very fast. It uses the 1/8 rate when the person is silent or nearly silent. In most areas the 13-K vocoder is the standard, but the originator of the call controls which vocoder is used.

The vocoder algorithm that is used is Coded Excited Linear Prediction (CELP). The actual specific algorithm used for cdmaOne is QCELP, a Qualcomm variant of CELP that assists in increasing the available system capacity.

Two rate-families are currently supported, the first based on 9600 bps and the second on 14 400 bps.

Advantages of CDMA

One of the main advantages claimed for CDMA is a vast improvement in the efficiency of the use of the spectrum. However, this parameter is very difficult to calculate in real terms, and there are many claims and counterclaims regarding the measurable advantages. This is because it depends to a large extent on a variety of aspects, including the level of interference from outside the cell, the cell density and such like. Accordingly, different methods of calculating the advantages yield different results. Despite this, some useful comparisons can be made with systems such as GSM. It is accepted that fifteen voice channels can be placed on each CDMA carrier, and as each carrier is 1.25 MHz wide (1.2288 MHz in exact terms), this represents a spectrum usage figure of twelve voice calls per MHz. GSM can carry eight voice channels per carrier and there are five carriers per MHz, which appears to provide a usage of forty calls per MHz. However, this is not the full story because neighbouring cells have to use different frequencies. With a cluster size of twelve (i.e. twelve cells in a cluster) this figure reduces to 3.3 calls per MHz, although when the frequency hopping facility of GSM is employed, cluster sizes of seven can be used and the frequency usage rises to six calls per MHz. Comparing the two systems, when other factors (such as handoff) that use up some of the capacity of a CDMA system are taken into account, it appears that cdmaOne has a capacity improvement of around 30 per cent. Whilst much smaller than the original figures seemed to indicate, it is still very important when the level of calls that can be carried represents the earnings that can be made by the telephone company, and a figure of 30 per cent is very significant.

Another advantage of CDMA is that new cells can be added relatively easily without the need for a considerable amount of frequency planning. By adding new cells further capacity is gained, albeit at the expense of additional interference because there are more mobiles and base stations that can interfere with one another. Although there is a loss of capacity in each cell, there are more cells and this brings an overall improvement in capacity – often by as much as fifteen additional channels for the addition of a new base station. Fortunately, by tailoring the antennas in each base station to provide the required coverage it is possible to limit the levels of overlap in the cells and thereby reduce some of the causes of interference.

CHAPTER NINE

CDMA2000

With IS-95 established and very successful, it became necessary to upgrade the system with a high-speed data capability to enable it to carry much faster data and thereby enhance the revenue growth. The approach adopted by those developing the CDMA technology from cdmaOne/IS-95 was to have a system in which there was a well-defined migration path forward but with full backward compatibility. As a result of this the IS-2000 standard was developed to enable the higher 3G data rates to be provided.

Within IS-2000, a number of further developments were included. It was envisaged that with many more areas moving towards 3G standards and the old AMPS systems being made obsolete, it would be possible to have systems operating on a wider bandwidth. As a result of this, the new standards allowed for systems that would use the single channel bandwidth (1X) and also ones that would use three times the bandwidth (3X).

The first stage of the 3G development in the form of CDMA2000 1X can double the voice capacity of cdmaOne networks and deliver peak packet data speeds of 307 kbps in mobile environments. CDMA2000 1X has been designated a 3G standard, and is widely deployed with operators in all continents of the world. Like cdmaOne, 1X is widely deployed in North America and the Asia Pacific regions as well as many other countries around the globe.

CDMA2000 1X is the basic 3G standard but, in what is termed CDMA2000 1xEv, there are further developments. The first of these, known as CDMA2000 1xEV-DO (evolution data only or, as it is becoming more widely known, EVolution Data Optimized) is something of a sideline from the main evolutionary development of the standard. It is defined under IS-856 rather than IS-2000 and, as the name indicates, it carries only data, but at speeds up to 3.1 Mbps in the forward direction and 1.8 Mbps in the reverse direction, the speed in the reverse link being upgraded as part of Release A of the standard. The first commercial CDMA2000 1xEV-DO network was deployed by SK Telecom (Korea) in January 2002 (see Figure 9.1).

The next logical evolution of the system is to incorporate both data and voice into the standard. This is exactly what CDMA2000 1xEV-DV achieves. This is catered for under Release C of the IS-2000 standard, and is effectively 1X with additional high-speed data channels. In this way it is able to provide complete backward compatibility with both CDMA2000 1X and cdmaOne. In addition to this, the migration requires comparatively few upgrades to a 1X system and as such

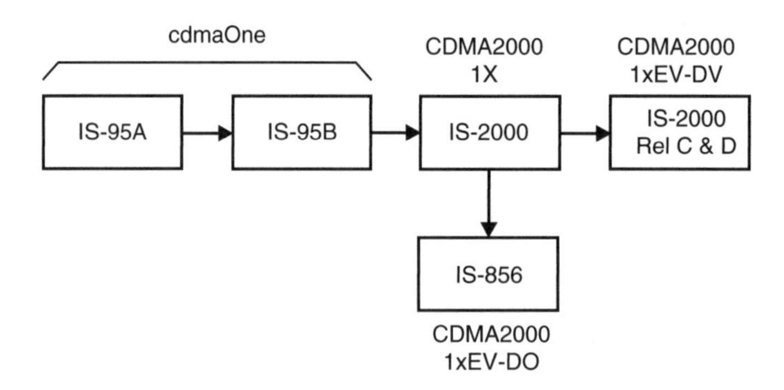

Figure 9.1 Development of CDMA2000.

it is a very attractive option for network operators. Further developments took place with Release D of the standard, which retained the peak data rates of 3.1 Mbps data in the forward link and increased the speed in the reverse link to 1.84 Mbps, as well as a number of other upgrades.

1X and 3X

When talking of CDMA2000, mention will often be made of 1XRTT and 3XRTT. These refer to two Radio Transmission Technologies. The original IS-95 and deployments of CDMA2000 utilize the 1.25-MHz channel spacing, which provides what is effectively the first phase of the 3G development and rollout. However, to enhance the performance beyond that possible using technologies such as 1xEV-DO and 1xEV-DV, the channel bandwidth of 1.25 MHz is insufficient. Accordingly, by increasing the bandwidth, higher data rates are possible. The further evolution of the CDMA2000 system involves utilizing channel bandwidths of three times the standard 1.25-MHz bandwidth under what is termed 3XRTT. Further bandwidth increases to 5X, 7X and so forth could in theory be contemplated.

For 1X technology spreading rate 1 (SR1) is used, where the signal is spread to occupy a bandwidth of 1.25 MHz. Here the spread rate is the same as that used for IS-95, i.e. 1.2288 Mcps. For 3X technology, spreading rate 3 (SR3) is used. Here, the spreading rate is 3.6864 Mcps. It is found that if the spreading rate remains the same but the data rate increases, as happens with video downloads and other 3G applications, the processing gain decreases. Accordingly, the coverage and signal strength need to be improved to match the new conditions. By increasing the spreading rate, the performance can be boosted without the need for improvements in coverage.

Radio configurations

The CDMA2000 standard offers a large degree of flexibility in terms of its configuration. Both the forward and reverse links can be configured differently to enable the optimum performance to be achieved for a given operator. It is possible to utilize different data rates, modulation and encoding, although the handsets must be compatible with the configuration to be able to operate correctly.

Table 9.1 Forward link radio configurations for CDMA2000 1X.

Radio configuration	Spreading rate	Data rates (bps)
1	1	1200, 2400, 4800, 9600
2	1	1800, 3600, 7200, 14.4k
3	1	1200, 1350, 1500, 2400, 2700, 4800, 9600, 19.2k, 38.4k, 76.8k, 153.6k
4	1	1200, 1350, 1500, 2400, 2700, 4800, 9600, 19.2k, 38.4k, 76.8k, 153.6k, 307.2k
5	1	1800, 3600, 7200, 14.4k, 28.8k, 57.6k, 115.2k, 230.4k
6	3	1200, 1350, 1500, 2400, 2700, 4800, 9600, 19.2k, 38.4k, 76.8k, 153.6k, 307.2k
7	3	1200, 1350, 1500, 2400, 2700, 4800, 9600, 19.2k, 38.4k, 76.8k, 153.6k, 307.2k, 614.4k
8	3	1800, 3600, 7200, 14.4k, 28.8k, 57.6k, 115.2k, 230.3k, 460.8k
9	3	1800, 3600, 7200, 14.4k, 28.8k, 57.6k, 115.2k, 230.4k, 460.8k, 518.4k, 1036.8k
		NB RC1 and 2 are for IS-95B

Table 9.2 Reverse link radio configurations for CDMA2000 1X.

Radio configuration	Spreading rate	Data rates (bps)
1	1	1200, 2400, 4800, 9600
2	1	1800, 3600, 7200, 14.4k
3	1	1200, 1350, 1500, 2400, 2700, 4800, 9600, 19.2k, 38.4k, 76.8k, 153.6k, 307.2k
4	1	1800, 3600, 7200, 14.4k, 28.8k, 57.6k, 115.2k, 230.4k
5	3	1200, 1350, 1500, 2400, 2700, 4800, 9600, 19.2k, 38.4k, 76.8k, 153.6k, 307.2k, 614.4k
6	3	1800, 3600, 7200, 14.4k, 28.8k, 57.6k, 115.2k, 230.4k, 460.8k, 1036.8k
		NB RC1 and 2 are for IS-95B

Accordingly, it is possible for the operator to choose which configurations are to be used in both the forward and reverse links. The different configurations are naturally called radio configurations, and they are defined in Tables 9.1 and 9.2.

CDMA2000 1X

A number of updates and changes were introduced to improve the performance of CDMA2000 1X IS-2000 over cdmaOne IS-95. However, in all cases backward compatibility has been maintained, allowing both IS-95 and IS-2000 mobiles to access the same base stations.

For IS-2000, CDMA2000 1X, several new methods of coding and spreading are used, and these enable much higher capacities to be achieved. The first major change is that the Walsh codes used are a different length. For IS-95, the length of the code was 64 bits and there was a choice of 64 codes per carrier. For 1X, the codes are generally 128 bits long, allowing for a total of 128 different codes. In addition to this, CDMA2000 uses stronger error coding functions and also turbo-codes rather than the convolutional codes used for IS-95. This enables higher-speed data to be sent. In addition to this, interleaving and symbol repetition are used to provide the various data rates.

Turbo-codes are a class of error correction codes that enable transfer rates over a noisy channel to approach the 'Shannon' limit. The turbo-coding principle was first proposed in 1993 by Professors Claude Berrou and Alain Glaxieux. Originally their claims that the codes could double throughput for a given power were treated with scepticism, but their findings were eventually proved to be accurate. Turbo-coders use powerful interleavers that reduce the susceptibility of a data stream to random and impulsive noise. By working on 'soft' bits from a radio receiver, the turbo-codes enable the decoder to extract the maximum level of data from the noisy signals. Turbo-codes require two encoders and two decoders per link, and these blocks operate in parallel and work synergistically. They also use an iterative process to reduce the amount of processing required, but despite this they still require more processing power than previous coding systems such as convolutional codes.

Apart from the improvements in the spreading and channel generation, there are also changes in the air interface itself. The IS-95 forward link uses a form of QPSK where the data on both the I and Q channels are the same. However, for CDMA2000 the I and Q channels are different, and this gives the advantage that half the bandwidth can be used for the same number of chips, or twice the number of chips can be sent in the same bandwidth. While this does make the reception more sensitive to phase errors, other improvements (including an improved system of forward power control and forward transmit diversity) enable these problems to be overcome to a sufficient degree, thereby effecting a doubling of the spectrum usage efficiency.

Similarly, there are significant changes on the reverse link where several new channels have been added. These include a pilot channel, as well as supplemental data channels and a control channel for signalling. Additionally, similar to the forward link, the reverse link now uses Walsh codes to differentiate between the different channels. A further change is that the format of the carrier modulation is changed. With the reverse link now transmitting multiple channels, the use of OQPSK would not prevent zero crossings. To achieve this, the modulation format is changed to a scheme known as Orthogonal Complex Quadrature Phase Shift Keying (OCQPSK). This form of modulation requires a number of stages. First the channels to be transmitted are split so that some take the I path and others the Q path, and then they are scrambled along with the Walsh code spreading. In the scrambling process the probability of zero crossings is identified and, using a scheme known as Orthogonal Variable Spreading Function (OVSF), the probability of zero crossings is reduced. Accordingly, the channels are spread with a Walsh code sequence and summed with the correct gain to produce the I and Q sequences. These are then further spread by a long PN code with its mobile-specific long mask to identify the mobile, and these I and Q sequences are modulated onto the carrier. Although particularly complicated, this form of modulation does have fewer zero crossings and the power amplifier in the mobile does not have to be run in a linear mode, thereby saving battery power.

Power control

For IS-95, fast power control was implemented on the reverse link to ensure that all the mobiles could be received at virtually the same power level. This is achieved by using a power control message every 1.25 ms. In this way, the 'near–far' problem could be overcome. However, on the forward link, where power control was not as critical, a power control message was sent every 20 ms to adjust the channel power to a given mobile.

While in many respects power control may not be quite as critical in the forward direction, when high data speeds are being carried accurate power control becomes more of an issue. Accordingly, a scheme similar to that implemented on the reverse link has been introduced on the forward link. It involves the use of two loops, as on the reverse link. The outer loop sets the required E_b/N_0 (this is effectively a measurement of signal-to-noise ratio, where it is the energy per bit divided by the noise power spectral density, to ensure a given acceptable frame error rate (FER). The inner loop then measures the received signal-to-noise ratio, and sends control bits back to the base station via the power control sub-channel on the new reverse pilot channel. In this way, power control is implemented every 1.25 ms as on the forward link.

Improvements have also been made to the reverse link power control. In view of the impact on high data-rate transmissions, the control can be implemented in 0.25-dB steps.

Beam formatting

The CDMA2000 system has been planned to accommodate smart antenna systems that allow beam formatting. Although advanced in concept, this technique has been in the minds of cellular and other wireless communications systems designers for many years. By adopting this technique, much smaller beams can be realized. This has the same effect as using smaller cells, and allows the re-use of the same codes more often. It also reduces the levels of interference, thereby allowing the full potential of the system to be realized and the highest data rates to be achieved.

In consequence, a number of new channels have been introduced to cater for this when it is implemented. For example, the primary pilot channel is used for the mobile to synchronize with the cell, and then an auxiliary pilot channel is used to allow it to synchronize to a specific beam.

Channels

CDMA2000 builds on IS-95 cdmaOne, and as a result it utilizes many of the channels that were found on the original system. However, in order to be able to provide the additional functionality and performance, a number of new channels have been incorporated. Channels are different for the forward and reverse links.

Forward link channels

1. *Forward Pilot Channel (F-PICH)*. This remains the same as it was on IS-95. It carries no data, and uses Walsh code 0. As Walsh code 0 is used, this is all zeros and therefore its length is immaterial; thus it retains compatibility with IS-95.

2. *Forward Transmit Diversity Pilot Channel (F-TDPICH).* While the F-PICH channel remains the primary pilot channel, this channel is used to provide a pilot for transmit diversity when two antennas are used. It provides a timing reference for the second antenna diversity signals. As F-PICH is the primary channel, this one is normally set at a power level below it.
3. *Forward Dedicated Auxiliary Pilot Channel (F-APICH).* This channel is used when smart antennas are employed. Under these circumstances a sector can be further divided into smaller sector beams, and this can be used to reduce the levels of interference. When this is implemented, each of these beams forms a new sector and needs its own pilot channel.
4. *Forward Auxiliary Transmit Diversity Pilot Channel (F-ATDPICH).* This channel is required when transmitting diversity in each beam when using a smart antenna system.
5. *Forward Sync Channel (F-SYNCH).* This channel provides the same functionality and operates in the same fashion as for IS-95. The timing of the channel is also aligned with that of the pilot channel, and as a result the mobile is able easily to decode the sync channel messages. Some additional information does need to be transmitted on this channel. For example, it is necessary to inform the mobile whether the base station is 1X or 3X compatible.
6. *Forward Paging Channel (F-PCH).* This channel retains the same structure as that used for IS-95, but a number of enhancements have been introduced. Information such as overhead and direction messages to specific mobiles can now be transmitted on other channels, and this improves the performance of the channel. Improvements have also been made to the sleep mode, so messages can be handled by a variety of channels when the mobile is not engaged in a call. These include the following:

   - Forward Quick Paging Channel (F-QPCH). This channel provides information to the mobile so that it can use slotted reception for its sleep mode. For IS-95, the mobile awoke at specific times to check the paging channel, which also carried other information. The forward quick paging channel is dedicated to this function, and is able to perform it more effectively. Using the paging channel in IS-95 the receiver had to wake up and receive a 96-ms slot, whereas using the F-QPCH it only needs to wake up and receive a 5-ms slot. This gives a significant reduction in the time for which the receiver needs to be awake, and considerably extends the standby battery life.
   - Forward Common Control Channel (F-CCCH). The F-CCCH is a forward link channel that transmits control information to a specific mobile. The frame sizes for this channel may be 5, 10 or 20 ms, and these provide data rates of 9.6, 19.2 and 38.4 kbps, respectively. When not on a call, mobile-specific messages can be sent on the F-CCCH.
   - Forward Broadcast Control Channel (F-BCCH). This channel is used to broadcast messages to all mobiles within the coverage area of the cell. These messages may include advertisements of news as well as static paging messages. The channel transmits data at 4.8, 9.6 or 19.2 kbps.

7. *Forward Common Power Control Channel (F-CPCCH).* Even when it is not involved in a call, the mobile still needs to receive bits for its power control. Each mobile therefore monitors a particular bit in this channel to enable it to adjust its power up or down by one increment.

To ensure the maximum speed, this information is not encoded. If any data errors are introduced, the error is quickly corrected by the next bit.

8. *Forward Fundamental Channel (F-FCH)*. This channel is still the main channel to carry the payload, whether voice or data. The fundamental channel uses a fixed amount of spreading, and the variable data rates are achieved by using symbol repetition.
9. *Forward Supplemental Code Channel (F-SCCH)*. The supplemental code channels are retained from IS-95 and contained within a traffic channel, and they are used to carry or data but may only be used with RC1 and RC2. For RC1 the frame size is 20 ms and the data rate is 9.6 kbps, whereas for RC2 the data rate is 14.4 kbps.
10. *Forward Supplemental Channel (F-SCH)*. This channel is contained within the traffic channel, and can only be used with radio configurations 3 to 5 for 1X and 6 to 9 for 3X. The channel can use variable length Walsh codes to provide a constant spreading rate. It provides for a wide range of data rates, from 1200 bps right up to 1 036 800 bps.
11. *Forward Dedicated Control Channel (F-DCCH)*. This channel carries information from the base station to the mobile. For IS-95 systems, the control information was sent on the forward traffic channel along with the voice data during a call. The introduction of this channel for CDMA2000 removes the requirement from the traffic channel to carry control information, thereby allowing the traffic channel to carry the data for which it was primarily intended and thus improving its efficiency.

Reverse channels

A number of improvements and new channels have been applied to the reverse link as well, including the following:

1. *Reverse Pilot Channel (R-PICH)*. This new channel for CDMA2000 provides a reference for base stations to gain initial access to a system and channel recovery. The channel also carries power control information for the forward link. Although the addition of this channel results in increased battery consumption, it provides significant advantages and has therefore been incorporated into the system.
2. *Reverse Access Channel (R-ACH)*. This channel operates in the same manner as that used for IS-95 and provides backward compatibility. It is used by the mobile to transmit random bursts when it attempts to access the system.
3. *Reverse Enhanced Access Channel (R-EACH)*. The reverse enhanced access channel performs the same functions as R-ACH, but has been enhanced to provide improved methods of access. There are two modes in which it can operate. The first is what is termed a basic access mode, where the channel carries access messages in a similar manner to that used on IS-95. The second is termed a reservation access mode, and it is used to reserve radio resources such as Internet access, etc. In both of these modes the channel is only used to establish the communications access. It can have frame sizes of 5, 10 and 20 ms, and provides data transmission at speeds of 9.6, 19.2 and 38.4 kbps.
4. *Reverse Common Control Channel (R-CCCH)*. This channel is an access channel that enables CDMA2000 to provide improved packet access. It provides the much faster access and lower

latency required for packet access. It offers a 20-ms 9.6-kbps channel, and in addition to this there are new 5- and 10-ms frames that offer data rates of 19.2 and 38.4 kbps, respectively. In operation, this channel is controlled by a closed-loop power control system.

5. *Reverse Dedicated Control Channel (R-DCCH)*. This is similar to the F-DCCH, and is used to carry signalling and user information in a call. The channel is only present if there are data to be transmitted.
6. *Reverse Fundamental Channel (R-FCH)*. This is the channel in the reverse link that is used to carry the voice and data payload. It allows for either 5- or 20-ms frames; the shorter frames are used to give lower latency (i.e. the phone gets the data it needs in 5 ms rather than 20 ms).
7. *Reverse Supplemental Code Channel (R-SCCH)*. This channel is the equivalent of the F-SCCH and is used with radio configurations 1 and 2. Again, the timing of this channel is offset from the fundamental channel by 1.25 mS. There can be up to seven used per traffic channel.
8. *Reverse Supplemental Channel (R-SCH)*. The reverse supplemental channel provides user data rates that vary from as low as 1200 bps up to 230.4 kbps (and 103.68 kbps for 3X). To accomplish this, the channel uses variable length Walsh codes. This channel applies only to radio configurations 3 and 4 for 1X, and 5 and 6 for 3X.

Packet data

With a growing level of data being sent, the packet data system has been adopted for this rather than circuit switched data, as it allows for far more efficient use of the data channels and therefore many users to access the capacity. This is particularly useful because much data is 'bursty' in nature, and circuit switching will tie up capacity even when no data are actually being sent. Accordingly, packet data can be sent easily over CDMA2000, where a slotted Aloha principle is used. Instead of using a fixed transmission power, it increases the transmission power for the random access burst after an unsuccessful access attempt. Then, after the mobile station has been allocated a traffic channel, it can transmit without scheduling up to a predefined bit rate. If the transmission rate exceeds the defined rate, a new access request has to be made.

When the mobile station stops transmitting, it releases the traffic channel but not the dedicated control channel. After a while it also releases the dedicated control channel, but maintains the link layer and network layer connections in order to shorten the channel set-up time when new data need to be transmitted. Short data bursts can be transmitted over a common traffic channel in which a simple ARQ is used to improve the error rate performance.

Handoff

One of the main limiting factors with IS-95 was the handoff process. With the ease with which soft handoffs could be performed, they occurred very frequently. This actually took up a considerable amount of the radio capacity, and in many instances it limited the capacity of the system. When developing CDMA2000, it was necessary to ensure that this problem did not occur again.

To achieve the required improvement, a number of measures were taken. A number of thresholds were introduced to determine whether a handoff is really necessary. These are based on elements such as the total pilot channel energy that is coherently demodulated. By utilizing the thresholds, the number of handoffs is reduced while still ensuring that handoffs are implemented when they are really required. By achieving a reduction in the number of handoffs that are undertaken, the system capacity that is available for real traffic is increased.

CDMA2000 1xEV-DV

While CDMA2000 1X was particularly successful, and the data rates offered did fall into some of the 3G classifications, it was nonetheless not the final evolution for 1X. Higher data rates were needed to retain users, and the IS-2000 standard was updated to accommodate this. As defined in Figure 9.1, it can be seen that Releases A and B of the standard catered for just the 1X capability, whereas the EV-DV capability was introduced in Release C and then upgraded in Release D.

The key features of CDMA2000 1xEV-DV are a high forward link capacity with data rates up to 3.1 Mbps peak with an average sector throughput of 1 Mbps. The original Release C version of the IS-2000 specification provided only for data rates of 384 kbps in the reverse direction, as it was anticipated that most of the data flow would be in the forward link. However, with applications such as video-conferencing becoming more likely, the Release D specification increased the maximum data rate in the reverse link to 1.8456 Mbps.

A further feature of the evolution was to provide concurrent support for both data and voice. This provides operators with an efficient use of their available spectrum as separate channels are not required for data and voice. The concept is that by taking advantage of the different usage patterns of voice and data, it is possible for the operator to make the most efficient use of the available spectrum. A further requirement of 1xEV-DV is that it retains backward compatibility with 1X and also with IS-95.

Despite the advantages of the scheme and its complete backward compatibility, many operators are adopting the 1xEV-DO solution for data, described below. Although this requires a separate 1X channel if voice traffic is to be carried (other than by Voice over IP), the data elements of the system have been optimized, and the industry appears to be moving toward the EV-DO standard. Nevertheless, an overview of the EV-DV solution is provided here.

New features

In order to meet the requirements for 1xEV-DV, a number of new features needed to be implemented. These included the addition of new channels and an adaptive modulation and coding scheme, and the addition of ARQ (Automatic Repeat reQuest – a mechanism used to trigger the retransmission of data packets received in error) to the physical layer and cell switching.

New channels

The EV-DV specification utilizes one new traffic channel and three new control channels as follows:

1. *Forward Packet Data Channel (F-PDCH)*. This is the main packet data channel. There can be two per sector, users being separated by both code-division and time-division techniques. The channel carries data, as well as some layer 3 signalling. To achieve the high data rate required, adaptive modulation and hybrid ARQ are applied.
2. *Forward Packet Data Control Channel (F-PDCCH)*. This channel is a forward link control channel, and it provides the mobile with the information it needs to correctly identify data intended for it on the F-PDCH. In more specific terms, the information carried is the MAC-ID, the F-PDCH packet size, the number of slots per sub-packet and the last Walsh code index. The MAC-ID is an 8-bit identifier known by the mobile and base station. It signifies the data being carried on the F-PDCH that is destined for the mobile. This identifier is set up as part of the call set-up procedure, and it remains the same during the duration of the call. The F-PDCH packet size simply indicates the size of the packet being transmitted, and the number of slots informs the mobile of the number of time division slots it occupies. The last Walsh code index is used by the mobile to determine the number of Walsh codes used for the data transmission on the F-PDCH to allow code division multiplexing on the channel.
3. *Reverse Channel Quality Indicator Channel (R-CQICH)*. This channel is used for the mobile to report the results of measurements of the channel quality of the best serving sector back to the base station.
4. *Reverse Acknowledgement Channel (R-ACKCH)*. This channel is used to inform the base station whether or not a packet transmitted on the F-PDCH has been successfully received.

Tables 9.3 and 9.4 show the channel types on forward and reverse CDMA links, respectively, for spreading rate 1.

Adaptive modulation

CDMA2000 1xEV-DV supports much higher levels of data throughput than does the 1X service. One of the key elements in being able to achieve this is the fact that adaptive modulation and coding is used. This is able to modify the RF transmission to best adapt to the environment and signal path that exists between the transmitter and the receiver (i.e. the base station and the mobile) or *vice versa*.

The adaptation is achieved by varying the RF packet duration, the number of bits per packet and the coding algorithm. The RF packet duration can be 1.25, 2.5 or 5 ms, and the number of bits per packet varies from 408 to 3864 bits. It will be seen that the variation in the number of bits per packet varies over a wider range than the packet length; this is achieved by also changing the modulation type, using QPSK, 8-PSK or 16 QAM. By making a choice from all these elements, it is possible to make the best utilization of the carrier.

Table 9.3 Channel types on forward CDMA link for spreading rate 1 (as IS-2000 Rel D/3GPP2 C.S0002).

Channel type	Maximum number
Forward pilot channel	1
Transmit diversity pilot channel	1
Auxiliary pilot channel	Not specified
Auxiliary transmit diversity pilot channel	Not specified
Sync channel	1
Paging channel	7
Broadcast control channel	8
Quick paging channel	3
Common power control channel	Not specified
Common assignment channel	7
Forward packet data control channel	2
Forward common control channel	7
Forward rate control channel	Not specified
Forward grant channel	Not specified
Forward acknowledgement channel	1
Forward dedicated control channel	Not specified
Forward fundamental channel	Not specified
Forward supplemental code channel (RC1 and RC2 only)	Not specified
Forward supplemental channel (RC3 through to RC5 only)	Not specified
Forward packet data channel	2

Table 9.4 Channel types per mobile station on reverse CDMA link for spreading rate 1 (As IS-2000 Rel D/3GPP2 C.S0002).

Channel type	Maximum number
Reverse pilot channel	1
Reverse secondary pilot channel	1
Access channel	1
Enhanced access channel	1
Reverse common control channel	1
Reverse packet data control channel	1
Reverse request channel	1
Reverse dedicated control channel	1
Reverse acknowledgement channel	1
Reverse channel quality indicator channel	1
Reverse fundamental channel	1
Reverse supplemental code channel (RC1 and RC2 only)	7
Reverse supplemental channel (RC3 and RC4 only)	2
Reverse packet data channel (RC7 only)	1

Hybrid automatic repeat request

In order to maintain the high data transmission rates of 1xEV-DV, it is necessary to ensure the fast re-transmission of any packets received with errors. As a result, the Automatic Repeat Request (ARQ) has been migrated from the MAC layer to the physical layer. Hybrid ARQ (HARQ) then improves the throughput by combining, rather than discarding, those packets that failed with the current attempt. In this way it is able more quickly and accurately to receive the data and re-build it if necessary. By adopting HARQ, which acts in the physical layer, shorter round trip delays are experienced compared to those using higher layer retransmission schemes employed in the Radio Link Protocol (RLP). This considerably reduces the possibility of data session timeout when using protocols such as TCP/IP. Additionally, it reduces latency, increases the average throughput and is more spectrally efficient.

Base station selection

When selecting a cell, the mobile will make RF quality measurements to enable it to connect to the base station providing the best RF link quality. Using a scheme known as cell selection, no soft handoff is undertaken for the EV-DV channels, although it is still active for the CDMA2000 1X traffic. By adopting this approach, the large overhead on the network resources caused by soft handoff is greatly reduced.

The system of cell selection requires that the base stations synchronize the F-PDCH data stream since it is possible that the mobile will have channels belonging to multiple base stations in its active set. When selecting a base station, the mobile indicates this selection via R-CQICH; the network receives this message and transmits the 1xEV-DV forward link data via the relevant base station.

Release D

The initial release of the IS-2000 specification defining 1xEV-DV appeared in the release C version of the specification. Further updates were undertaken to increase the speed of the reverse link to provide data at a much increased speed, as well as providing some improvements for the forward link as well.

To achieve the increased reverse link data transfer rates, new channels were introduced as follows:

1. *Reverse Packet Data Channel (R-PDCH)*. This is a data bearing channel in the reverse link; it has a fixed duration of 10 ms and it can only transmit any of the fixed packet sizes of length 192, 408, 792, 1560, 3096, 4632, 6168, 9240, 12 312, 15 384 or 18 432 bits. Using the variable packet sizes, four channel HARQ and a maximum of three transmissions enable data rates from 6.4 kbps right up to 1.8432 Mbps to be achieved.
2. *Reverse Packet Data Control Channel (R-PDCCH)*. This channel was introduced to avoid the problem of blind detection and decoding at the base station, and the channel carries information relating to the packet format on the R-PDCH.
3. *Reverse Request Channel (R-REQCH)*. This channel supports the autonomous transmission mode, permitting the mobile to start transmission at any given time up to a pre-authorized

maximum data rate. The autonomous mode allows the mobile to start up a service at any time while simultaneously reducing the delay. In this way the system is able to provide different levels of quality of service, to enable it to meet the needs of the mobile more accurately.

4. *Reverse Secondary Pilot Channel (R-SPICH)*. This channel serves three purposes: to facilitate fast and simple detection, to improve reverse link power control, and to improve the coherent demodulation.

Further channels were also introduced in the forward link:

1. *Forward Grant Channel (F-GCH)*. This channel is monitored by all mobiles within the sector served by the channel. It is used when high-speed data transmissions are to be undertaken, and it allows a mobile access to the network for data transmission above a given rate.
2. *Forward Indicator Control Channel (F-ICCH)*. This was added to facilitate rate control on a per user or per sector basis.
3. *Forward Acknowledgement Channel (F-ACKCH)*. This channel provides synchronous acknowledgements to the received reverse link data packet transmissions. The channel is capable of serving up to 192 users in any given sector.

Broadcast and multicast services

In addition to the fast data in both forward and reverse links, EV-DV is also able to support broadcast multimedia transmissions from a single source to multiple users with users in either an active or an idle state. This is achieved by enhancing the forward supplemental channel and the forward fundamental channel, and this can be achieved in one of three ways.

In the first approach, the F-SCH channel is shared among the idle mobile stations and no reverse link channels are required. To enable the service to operate, new Broadcast and Multicast Services (BCMCS) specific upper layer messages are sent to the mobiles with parameters for configuration information. Handoff is autonomous, the mobile selecting the best base station from which to receive the data.

For the second type of BCMCS, the F-FCH is shared among the active mobiles. When a mobile is in a traffic state, the F-FCH can be shared and used to carry BCMCS content. For this type, the F-CPCCH carries the power control command and the F-DCCH carries the individual signalling, using a TDM approach for each mobile. Each mobile has its own long code mask for reverse link channelization. Using this mode, data rates of 9.6 and 14.4 kbps are supported. The advantage of this type of BCMCS is that it provides dynamic coverage and optimizes base station power usage and code usage. Additionally, it allows soft handoff.

The third type of BCMCS shares the F-SCH among the active mobile stations, and it can be used by mobiles in a traffic state. The F-SCH is shared by all users. Signalling procedures remain the same for the shared F-SCH.

Fast call set-up

With any cellular communications system, call set-up can take a while. For CDMA2000, it can take up to 12 seconds under worst-case scenarios. This length of time is unacceptable for applications such as Push to talk over Cellular (PoC). Much of the delay arises from the procedures that have to be undertaken. A call set-up can be split into three stages: paging, access, and channel set-up and service negotiation.

To reduce the time for fast call set-up, a number of measures have been taken. The first is to reduce the paging delay. When an application such as PoC is encountered, the mobile may enter a reduced slotted or a non-slotted mode for a period of time. This decreases the time between paging slots, and may reduce the paging time by several seconds. Additionally, the base station may perform channel assignment without the access procedure, again saving several seconds. Finally, the traffic channel set-up has been simplified, thereby saving time. Combined, these measures enable very fast call set-up to be achieved when it is required.

Mobile equipment identifier

Mobile phones are assigned a 32-bit Electronic Serial Number (ESN). However, with the enormous growth in mobile phone production these numbers are running out. To resolve the problem, a 56-bit identifier known as the Mobile Equipment Identifier (MEID) has been introduced. Careful implementation has been required to ensure that both ESNs and MEIDs can operate alongside each other. Having overcome these problems, this has enabled a much greater number of identification serial numbers to be used, preventing the shortage that was due to occur.

CDMA2000 1xEV-DO

Despite the fact that EV-DO may be considered something of a sidetrack from the main IS-2000 development, it has nevertheless gained significant acceptance and now appears to be the main route to provide 3G data speeds. As the name implies, it has been optimized for data transmission, the letters now standing for 'data optimized' rather than the initial meaning of 'data only'. However, being a data-only system, channels that carry DO transmissions may not carry voice. A separate 1X carrier is required for voice transmissions unless they are carried, for example, as data in the form of Voice over IP (VoIP).

The CDMA2000 1xEV-DO cell phone system is defined under IS-856 rather than IS-2000, which defines the other CDMA2000 standards. In Release 0 of the standard, the maximum data rate was 2.4576 Mbps in the forward (downlink) with 153.6 kbps in the reverse (uplink) direction – the same as CDMA2000 1X. However, in a later release of the standard, Release A, the forward data rate has risen to 3.072 Mbps and the reverse data rate to 1.8432 Mbps (see Figure 9.2).

The forward channel forms a dedicated variable-rate packet data channel with signalling and control time multiplexed into it. The channel is itself time-divided and allocated to each user on a demand and opportunity driven basis. A data-only format was adopted to enable the standard to be

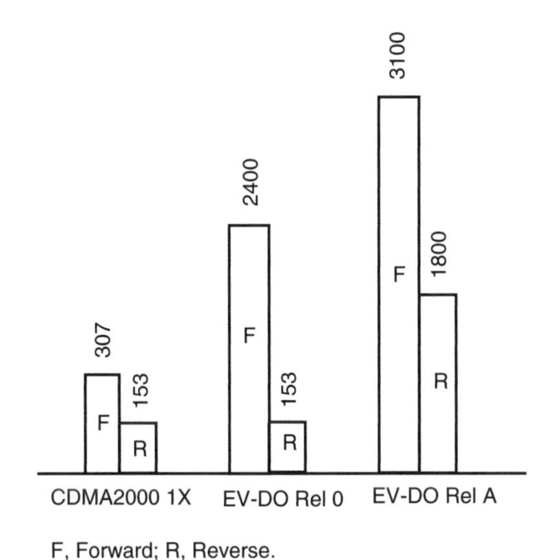

Figure 9.2 Data rates for CDMA2000 1xEV-DO.

optimized for data applications. If voice is required, then a dual mode phone using a separate 1X channel for the voice call is needed. In fact, the 'phones' used for data only applications are referred to as Access Terminals (ATs).

EV-DO air interface

The EV-DO RF transmission is very similar to that of a CDMA2000 1X transmission. It has the same final spread rate of 1.2288 Mcps, and it has the same modulation bandwidth because the same digital filter is used. Although 1xEV-DO transmissions have many similarities with 1X transmissions, they do not occupy the same channels simultaneously, and therefore EV-DO uses dedicated paired channels for its operation.

Forward link

The forward link possesses many features that are specific to EV-DO, having been optimized for data transmission, particularly in the downlink direction. Average continuous rates of 600 kbps per sector are possible. This is a six-fold increase over CDMA2000 1X, and is provided largely by the ability of 1xEV-DO to negotiate increased data rates for individual ATs because only one user is served at a time.

The forward link is always transmitted at full power and uses a data rate control scheme rather than the power control scheme used with 1X, and the data is time-division multiplexed so that only one AT is served at a time.

In order to be able to receive data, each EV-DO AT measures signal-to-noise ratio (S/N) on the forward link pilot in every slot (i.e. 1.667 ms). Based on the information this provides, the AT sends a data rate request to the base station or Access Node (AN). The AN receives requests from a variety of ATs, and decisions have to be made regarding which ATs are to be served next. The AN endeavours to achieve the best data transfer, and this is done by serving those ATs offering a good signal-to-noise ratio. This is achieved at the expense of users experiencing poor propagation conditions, which will typically be those at some distance from the AN's antenna.

As with 1X, accurate time synchronization is required between the EV-DO access nodes. To achieve this, time information is taken from the Global Positioning System, as this is able to provide an exceedingly accurate time signal.

Forward link channels

A number of channels are transmitted in the forward direction to enable signalling, data and other capabilities to be handled. These channels include the traffic, MAC, control and pilot channels. These are time-division multiplexed.

1. *Traffic channel.* This channel uses quadrature phase shift keying (QPSK) modulation for data rates up to 1.2288 Mbps. For higher data rates higher-order modulation techniques are used, in the form of 8PSK with 3 bits per symbol or 16QAM with 4 bits per symbol. The levels of the I and Q symbols are chosen so that the average power remains the same regardless of the modulation scheme employed. The incoming data to be used in the modulation come from the turbo-coder, and are scrambled by mixing with a pseudorandom number (PN) sequence. The initial state of the PN is derived from known parameters, and is unique for each user. Every packet starts at the same initial value of the PN sequence. At the beginning of the transmission to each user, there is a preamble that contains the user ID for the data. Its repeat rate is determined by the data rate because lower data rates require higher repeat values. However, even at its largest the preamble will fill no more than half the first slot.
2. *Control channel.* This channel carries the signalling and overhead messages. It has the same structure as the traffic channel.
3. *Pilot channel.* The differentiator between the cell and the sector is still the PN offset of the pilot channel, and the pilot signal is only gated on for 192 chips per slot. The pilot PN sequence is the time-reversed sequence of 1X. This is done so that mobiles do not confuse 1X transmissions with EV-DO ones.
4. *Medium Access Control (MAC) channel.* This channel carries a number of controls, including the Reverse Power Control (RPC), the Data Rate Control (DRC) Lock and the reverse activity (RA) channels.

Reverse link

The reverse link for 1xEV-DO has a structure similar to that of CDMA2000. In EV-DO all signalling is performed on the data channel, and this means that there is no dedicated control channel. The data channel can support five data rates, which are separated in powers of 2 from 9.6 to 153.6 kbps. These rates are achieved by varying the repeat factor. The highest rate uses a turbo-coder with lower gain. The following channels are transmitted in addition to those used with 1X:

1. *Reverse Rate Indicator (RRI) channel*. This indicates the data rate of the reverse data channel.
2. *Acknowledgement (Ack) channel*. This channel is transmitted after the AT detects a slot, with the preamble identifying it as the recipient of the data.
3. *Data Rate Control (DRC) channel*. This channel contains a 4-bit word in each slot to allow the choice of twelve different transmission rates. The phone also uses a 3-bit DRC cover to identify the base station from which it wants to receive packets on the forward traffic channel.

Mobile IP

One of the advantages of 3G high-speed data networks is their ability to allow users to connect their computers to networks. People need to take laptop computers with them and use them anywhere, as if they were working from their 'home' or normal network. While it is possible to make connections reasonably easily, improvements are being put in place to ensure full mobility and ease of use.

As infrastructures and standards are already in place for data transfer, it is necessary to adapt them to take account of mobility rather than introducing completely new techniques. The most common services are the data services using the Internet Protocol (IP). When using this, a user (which may be any form of node or computer) is normally connected to a particular network or sub-network. Moving the computer from one network or sub-network to another creates problems because routing tables need to be updated to enable the data to reach the user at the new location.

Under normal circumstances, users are attached to what is termed their home network. This is the network to which the computer is usually attached, and all the routing tables are set up for the computer successfully to send and receive data from this location. Using their home network IP address, users can move anywhere within this particular network with no problem.

The most common example of operating in a 'home' environment is when a computer connects to its normal wired or wireless LAN.

If a node or computer needs to move outside its home network, to what is termed a 'foreign' network, then it needs a means of connecting back to the home network so that data packets sent to the home network can be forwarded to the new location and *vice versa*. This is done by using a 'foreign agent' (FA). Each network has its own foreign agent to enable the provision of mobile data operation. It operates by advertising its presence and services on its network, and looking for any foreign users that may have attached to its network. Once a foreign user is found, it communicates with that user to establish the required information to link to the home network.

Similarly, on the home network there is an equivalent agent called the 'home agent' (HA). The home agent acts as what is termed a 'proxy' for the mobile user – in other words, it takes the place of the home IP location and routes data to the foreign agent, allowing communication with what is termed the 'correspondent node' (CN).

In operation, the foreign agent connects to the home agent when authentication is complete, and it uses an IP tunnel for communication. In this tunnel, IP packets are packed within IP packets communicating the data (Figure 9.3).

Mobile networks are also starting to employ mobile IP. Work is well advanced on the CDMA2000 system, and it will also be implemented on UMTS, being included only in later releases of that standard.

The way in which IP is used on a cellular system is very similar to a dial-up phone connection where a computer is to connect to the Internet. Here, the user makes a connection using what is termed the Point to Point Protocol (PPP). As the connection is established, the service provider assigns an IP address to the user. Once this has occurred, the data packets have an address to which they can be routed. While the connection is maintained, all packets of data are routed to this IP address and others are obviously sourced from it.

The same happens when a mobile phone connects to the Internet. A connection is established and an IP address assigned to the phone or laptop. This works well while the phone is connected to the same base station or local switching centre. However, when it needs to move away a problem arises, because each switching centre acts as a different sub-network. When a mobile moves from one switching centre to the next, the connection needs to be broken and a new one established using a new IP address. For CDMA2000 networks this is known as Simple IP. This is clearly not an efficient method of operating, and considerably reduces the performance of the system because it breaks all the IP-based connections made by applications running on the mobile node.

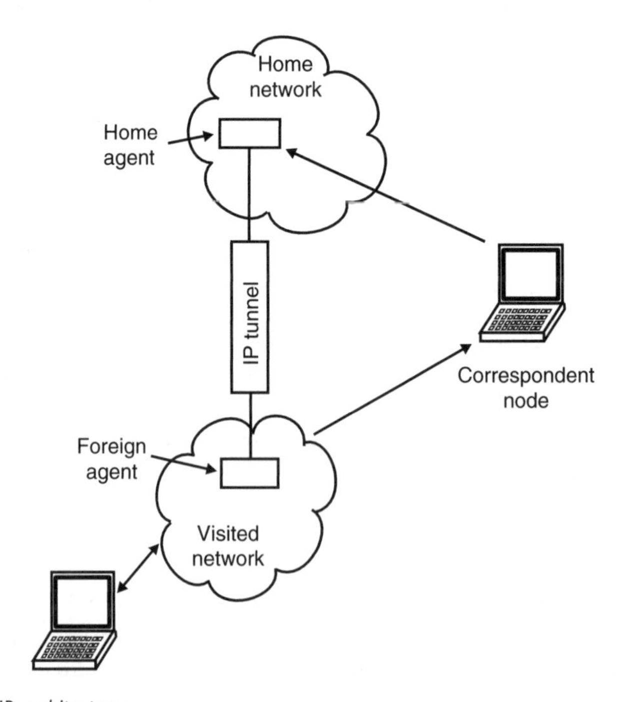

Figure 9.3 Mobile IP architecture.

Accordingly, the mobile phone system is treated as a network in the same way as a wired network. Each switching centre has a foreign agent, which operates in the same way that it does for a wired LAN system. It communicates messages from a operating mobile that has moved away from its home switching centre, and in this way the IP connection is not broken.

By adopting this approach the foreign agents serving different switching centres are used, and the information is updated with the home agent as the mobile moves from one switching centre to the next. Although this complicates the handover process it enables a continuous connection to be maintained, despite the mobile moving its location and requiring to be served by different switching centres.

With the telecommunications scene changing rapidly, moving from a voice-centred service to a data-centred service and the hybrid approaches being offered to provide the optimum service, Mobile IP is clearly going to become an increasingly important technique to be used to enable seamless transition from one area to the next, and one technology to the next.

UMTS

UMTS, the Universal Mobile Telecommunications System, is the third-generation (3G) successor to the second-generation GSM-based technologies, including GPRS, and EDGE. Although UMTS uses a totally different air interface, the core network elements have been migrating towards the UMTS requirements with the introduction of GPRS and EDGE. In this way, the transition from GSM to UMTS does not require such a large instantaneous investment.

UMTS, which uses wideband CDMA (W-CDMA), has had a long history. Even as the first 2G systems were first being rolled out, it was clear that these would not cater for the demand forever. New technologies capable of providing new services and facilities would be required. With this in mind, the World Administrative Radio Conference started to reserve spectrum allocations for a new service at its meetings in 1992.

The next stage in the development arose when the International Telecommunications Union began defining a system, and the International Mobile Telecommunications System 2000 (IMT2000) started to take shape. In order to manage the new standard, a group known as the Third Generation Partnership Programme (3GPP) was formed. In fact, 3GPP is a global co-operation between six Organizational Partners (ARIB, CCSA, ETSI, ATIS, TTA and TTC) who are recognized as being the world's major standardization bodies from Japan, China, Europe, the USA and Korea. The establishment of 3GPP was formalized in December 1998 by the signing of The Third Generation Partnership Project Agreement.

The original scope of 3GPP was to produce globally applicable technical specifications and technical reports for a third-generation mobile telecommunications system. This would be based upon the GSM core networks and the radio access technologies that they support (i.e. Universal Terrestrial Radio Access (UTRA), both Frequency Division Duplex (FDD) and Time Division Duplex (TDD) modes). Later the scope of 3GPP was increased to include the maintenance and development of the GSM Technical Specifications and Technical Reports, including its derivatives of GPRS and EDGE.

In view of the fact that UMTS is a 3G technology and is a successor to GSM with a defined migration path, some are now referring to it as 3GSM.

While 3GPP undertook the management of the UMTS standard, a similar committee was needed to oversee the development of the other major 3G standard, namely CDMA2000. This committee took on the name 3GPP2, and the standards bodies that were represented included ARIB, TTA, CWTS and TIA.

Capabilities

UMTS uses W-CDMA as the radio transmission standard. It employs a 5-MHz channel bandwidth (wider than the cdmaOne/CDMA2000 1XRTT channel bandwidth of 1.25 MHz), and as such it has the capacity to carry over 100 simultaneous voice calls, or to carry data at speeds up to 2 Mbps in its original format. However, with the later enhancements included in later releases of the standard, new technologies (including HSDPA and HSUPA) have enabled data transmission speeds of 14.4 Mbps.

Many of the ideas that were incorporated into GSM have been carried over and enhanced for UMTS. Elements such as the SIM have been transformed into a far more powerful USIM (universal SIM). In addition to this, the network has been designed so that the enhancements employed for GPRS and EDGE can be used for UMTS. In this way, the investment required is kept to a minimum.

A new introduction for UMTS is that there are specifications that allow both FDD and TDD modes. The first modes to be employed are FDD modes, where the uplink and downlink are on different frequencies. The spacing between them is 190 MHz for the Band 1 networks being currently used and rolled out. However, the TDD mode where the uplink and downlink are split in time with the base stations and then the mobiles transmitting alternately on the same frequency is particularly suited to a variety of applications. An obvious example is where spectrum is limited and paired bands suitably spaced are not available. It also performs well where small cells are to be used. As a guard time is required between transmit and receive, this will be less when transit times are smaller as a result of the shorter distances being covered. A further advantage arises from the fact that more data are carried in the downlink as a result of Internet surfing, video downloads and the like. This means that it is often better to allocate more capacity to the downlink. Where paired spectrum is used this is not possible; however, when a TDD system is used it is possible to alter the balance between downlink and uplink transmissions to accommodate this imbalance and thereby improve the efficiency. In this way, TDD systems can be highly efficient when used in picocells for carrying Internet data. The TDD systems have not been widely deployed, but this may occur more in the future. In view of its character, it is often referred to as TD-CDMA (Time Division CDMA).

System architecture overview

Like GSM, the network for UMTS can be split into three main constituents. These are the mobile station, called the User Equipment or UE, the base station sub-system, known as the Radio Network Sub-system (RNS), and the core network.

User equipment

The user equipment is very much like the mobile equipment used within GSM. It consists of a variety of different elements, including RF circuitry, processing, antenna, battery and the like. The UMTS/W-CDMA standard, with the many additional features and applications that can run, requires far greater levels of processing than were needed for GSM, GPRS and EDGE. Accordingly, the development programmes took longer than expected and the first phones appeared somewhat later than had been hoped. The levels of integration required were much greater than those needed previously, and this also resulted in higher drain on the battery. These problems had to be overcome to ensure that users were not disappointed and found the battery life of the new 3G phones much shorter than those of the old 2G and 2.5G phones. In addition to this, many new integrated circuits combined both analogue RF and digital areas of the phone to reduce the component count in the phones.

Another problem that had to be overcome resulted from the form of modulation used. It required that a linear RF amplifier was used for the transmitter, unlike the GMSK modulation used for GSM. The use of a linear RF amplifier increased the drain on the battery. To help overcome this, a number of novel RF design techniques (which are not within the scope of this book) were employed by designers to reduce the current drain, thereby improving the battery life.

Use was also made of the improved battery technology that was coming available. Many older phones had used nickel cadmium technology, but as batteries improved a move was made to nickel metal hydride types. Lithium ion (Li-ion) batteries then became standard, and they provided considerably greater energy storage densities, allowing phones to remain small and relatively lightweight while still retaining or even improving the overall life between charges.

The UE also contains a SIM card, although in the case of UMTS it is termed a USIM (Universal Subscriber Identity Module). This is a more advanced version of the SIM card used in GSM and other systems, but embodies the same types of information. It contains the International Mobile Subscriber Identity number (IMSI) as well as the Mobile Station International ISDN Number (MSISDN). The USIM may store more than one MSISDN, enabling the phone to be used for a number of different 'lines' – home, business, etc. The USIM is also programmed with an access priority control class. These classes are used to give different users different levels of priority when accessing the system. Levels 0–9 are assigned to regular customers, while 10–14 are assigned to users, such as the police and other safety personnel, who may need to gain access even when the system might otherwise be nearing capacity. Other information that the USIM holds includes the preferred language, to enable the correct language information to be displayed, especially when roaming, and a list of preferred and prohibited Public Land Mobile Networks (PLMN).

The USIM also contains a short message storage area that allows messages to stay with the user even when the phone is changed. Similarly, 'phone book' numbers and call information regarding the numbers of incoming and outgoing calls are stored.

The USIM is an integrated circuit assembly which contains the USIM Integrated Circuit Card (UICC). The information stored is in the form of Elementary Files (EF). The UICC has eight contacts available. The supply voltage is applied to pin 1, pin 2 is used as the reset line, and a clock reference is supplied on pin 3. Pin 7 is used for the data input and output line. The UICC may come in one of three varieties using 5 V, 3 V or 1.8 V. The UICC reader starts by applying the lowest voltage, and if no answer is received then the next higher voltage is applied. The USIM card is the same physical size as that used for GSM phones, as well as being used for some satellite applications, and complies with an ISO standard.

Radio network sub-system

This is the section of the network that interfaces to both the UE and the core network. It contains what are roughly equivalent to the Base Transceiver Station (BTS) and the Base Station Controller (BSC). Under UMTS terminology, the radio transceiver is known as the node B. This communicates with the various UEs, and with the Radio Network Controller (RNC). This is undertaken over an interface known as the Iub. The overall radio access network is known as the UMTS Radio Access Network (UTRAN; Figure 10.1).

The RNC component of the Radio Access Network (RAN) connects to the core network. The core network used for UMTS is based upon the combination of the circuit switched elements used

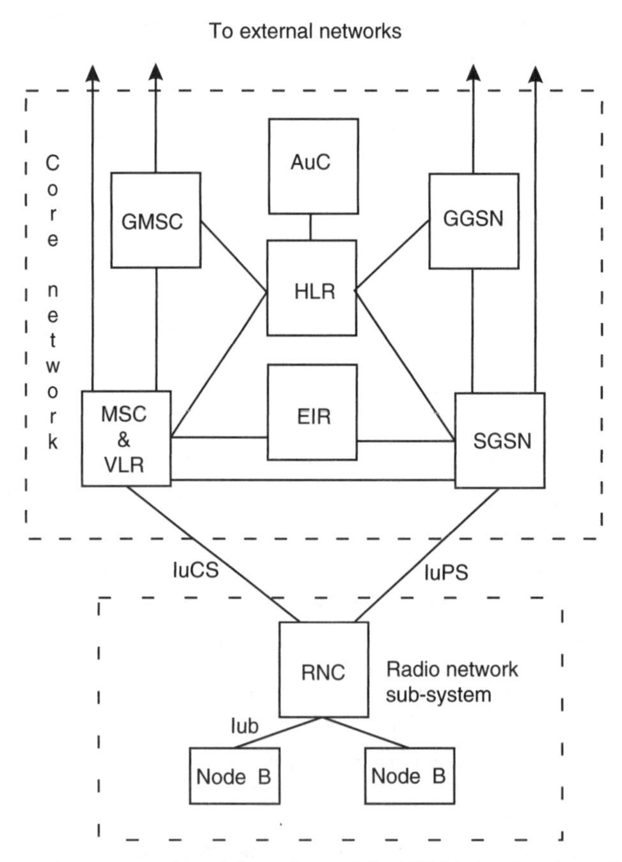

Figure 10.1 Diagrammatic representation of the makeup of the UTRAN showing the major entities.

for GSM plus the packet switched elements that are used for GPRS and EDGE. Thus the core network is divided into circuit switched and packet switched domains. Some of the circuit switched elements are Mobile services Switching Centre (MSC), Visitor Location Register (VLR) and Gateway MSC. Packet switched elements are Serving GPRS Support Node (SGSN) and Gateway GPRS Support Node (GGSN). Some network elements, like EIR, HLR, VLR and AUC, mentioned in Chapters 4 and 6, are shared by both domains and operate in the same manner that they did with GSM.

The Asynchronous Transfer Mode (ATM) is specified for UMTS core transmission.

The architecture of the core network may change when new services and features are introduced. The Number Portability DataBase (NPDB) will be used to enable subscribers to change network provider while keeping their old phone number. The Gateway Location Register (GLR) may be used to optimize the subscriber handling between network boundaries. MSC, VLR and SGSN can merge to become a UMTS MSC.

Protocols

The signalling protocol used in UMTS networks is an evolution of that used in GSM/GPRS networks. The protocol has been subdivided into the Access Stratum and Non-Access Stratum. The non-access stratum is responsible for mobility management and call control functions and, as its name implies, is independent of the radio access network type, allowing the signalling to be kept as common as possible between GSM/GPRS and UMTS networks. The access stratum consists of the layer 2 and lower part of the layer 3. The layer 3 part of the access stratum consists of the RRC, which corresponds to the RR sub-layer in GSM networks, and is responsible for control and allocation of radio resources. The layer 2, responsible for maintaining an appropriate link and supplying data to layer 1, has been divided into two sub-layers, MAC and RLC.

A further subdivision applied to the protocol is between the control plane (C-plane) and user plane (U-plane), where the C-plane encompasses the protocols used on the control channels and the U-plane covers traffic channel management.

Air interface

There are currently six bands that are specified for use, and operation on other frequencies is not precluded. However, much of the focus is currently on frequency allocations around 2 GHz. At the World Administrative Radio Conference in 1992, the bands 1885–2025 and 2110–2200 MHz were set aside for use on a worldwide basis by administrations wishing to implement International Mobile Telecommunications-2000 (IMT-2000). This does not, however, preclude its use for other applications.

Within these bands, the portions have been reserved for different uses:

Bands (MHz)	Use	Characteristics
1920–1980 and 2110–2170	Frequency Division Duplex (FDD, W-CDMA)	Paired uplink and downlink, channel spacing 5 MHz, raster 200 kHz. An operator needs three or four channels (2 × 15 MHz or 2 × 20 MHz) to be able to build a high-speed, high-capacity network
1900–1920 and 2010–2025	Time Division Duplex (TDD, TD/CDMA)	Unpaired, channel spacing 5 MHz, raster 200 kHz. Transmit and receive transmissions are not separated in frequency
1980–2010 and 2170–2200	Satellite uplink and downlink	

Carrier frequencies are designated by a UTRA Absolute Radio Frequency Channel Number (UARFCN). This can be calculated as:

$$\text{UARFCN} = 5 \times (\text{frequency in MHz}).$$

UMTS uses wideband CDMA as the radio transport mechanism. The channels are spaced by 5 MHz. The modulation that is used is different on the uplink and downlink. The downlink uses quadrature phase shift keying (QPSK) for all transport channels. However, the uplink uses two separate channels so that the cycling of the transmitter on and off does not cause interference on the audio lines – a problem that was experienced on GSM. The dual channels (dual channel phase shift keying) are achieved by applying the coded user data to the I or in-phase input to the DQPSK modulator, and control data which has been encoded using a different code to the Q or quadrature input to the modulator.

Spreading

As with any CDMA-based system, the data to be transmitted are encoded using a spreading code particular to a given user. In that way only the desired recipient is able to correlate and decode the signal, all other signals appearing as noise.

As described in Chapter 3, the data of a CDMA signal are multiplied with a chip or spreading code to increase the bandwidth of the signal. For W-CDMA, each physical channel is spread with a unique and variable spreading sequence. The overall degree of spreading varies to enable the final signal to fill the required channel bandwidth. As the input data rate may vary from one application to the next, so the degree of spreading needs to be varied accordingly.

For the downlink, the symbol rate that is transmitted is 3.84 Msps. As the form of modulation used is QPSK, this enables 2 bits of information to be transmitted for every symbol, thereby enabling a maximum data rate of twice the symbol rate, or 7.68 Mbps. Therefore, if the actual rate of the data to be transmitted is 15 kbps, then a spreading factor of 512 is required to bring the signal up to the required chip rate for transmission in the required bandwidth. If the data to be carried have a higher

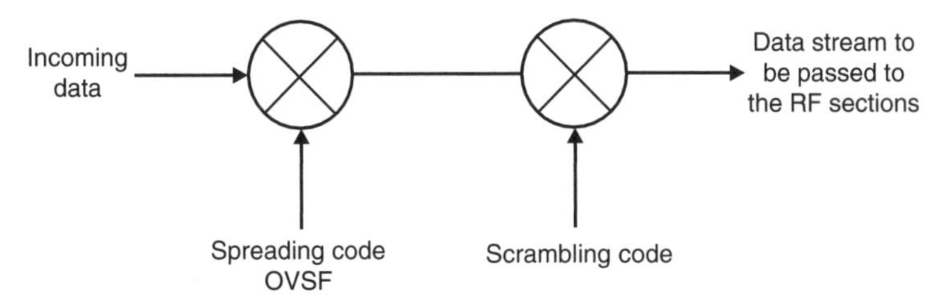

Figure 10.2 Spreading and scrambling process.

data rate, then a lower spreading rate is required to balance this out. It is worth remembering that altering the chip rate does alter the processing gain of the overall system, and this also needs to be accommodated in the signal processing. Higher spreading factors are more easily correlated by the receiver, and therefore a lower transmit power can be used for the same symbol error rate.

The codes required to spread the signal must be orthogonal if they are to enable multiple users and channels to operate without mutual interference. The codes used in W-CDMA are Orthogonal Variable Spreading Factor (OVSF) codes, and they must remain synchronous to operate. As it is not possible to retain exact synchronization for this, a second set of scrambling codes is used to ensure that interference does not result (Figure 10.2). This scrambling code is a pseudorandom number (PN) code. Thus there are two stages of spreading: the first using the OSVF code, and the second using a scrambling PN code. These codes are used to provide different levels of separation. The OVSF spreading codes are used to identify the user services in the uplink and the user channels in the downlink, whereas the PN code is used to identify the individual node B or UE.

On the uplink, there is a choice of millions of different PN codes. These are processed to include a masked individual code to identify the UE. As a result, there are more than sufficient codes to accommodate the number of different UEs likely to access a network. For the downlink, a short code is used. There are a total of 512 different codes that can be used, one of which will be assigned to each node B.

Synchronization

The level of synchronization required for the W-CDMA system to operate is provided by the Primary Synchronization Channel (P-SCH) and the Secondary Synchronization Channel (S-SCH). These channels are treated in a different manner to the normal channels, and as a result they are not spread using the OVSFs and PN codes but rather by using synchronization codes. Two types are employed: the first is called the primary code and is used on the P-SCH, while the second is named a secondary code and is used on the S-SCH.

The primary code is the same for all cells, and is a 256-chip sequence that is transmitted during the first 256 chips of each time slot. This allows the UE to synchronize with the base station for the time slot.

Once the UE has gained time slot synchronization, it knows the start and stop of the time slot but it does not have information about the particular time slot, or the frame. This is gained using the secondary synchronization codes.

There are sixteen different secondary synchronization codes in total. One code is sent at the beginning of the time slot, i.e. the first 256 chips. It consists of fifteen synchronization codes, and there are sixty-four different scrambling code groups. When received, the UE is able to determine before which synchronization code the overall frame begins. In this way, the UE is able to gain complete synchronization.

The scrambling codes in the S-SCH also enable the UE to identify which scrambling code is being used, and hence it can identify the base station. The scrambling codes are divided into sixty-four code groups, each having eight codes. This means that after achieving frame synchronization, the UE only has a choice of one in eight codes and it can therefore try to decode the CPICH channel. Once it has achieved this, it is able to read the BCH information and attain better timing, and to monitor the P-CCPCH.

Power control

As with any CDMA system, it is essential that the base station receives all the UEs at approximately the same power level. If not, the UEs that are further away and are lower in strength will not be heard by the node B, and only those that are close will be able to access the system. This effect is often referred to as the near–far effect.

It is also important for node Bs to control their power levels effectively. As the signals transmitted by the different node Bs are not orthogonal to one another, it is possible that signals from different ones will interfere. Accordingly, their power is also kept to the minimum required by the UEs being served.

To achieve the power control, two techniques are employed: open loop and closed loop. The open-loop technique is used during the initial access, before communication between the UE and node B has been fully established. The technique simply operates by making a measurement of the received signal strength and thereby estimating the transmitter power required. As the transmit and receive frequencies are different, the path losses in either direction will vary and therefore this method cannot be any more than a good estimate.

Once the UE has accessed the system and is in communication with the node B, closed-loop techniques are used. A measurement of the signal strength is taken in each time slot and, as a result of this, a power control bit is sent requesting the power to be stepped up or down. This process is undertaken on both the up- and downlinks. The fact that only one bit is assigned to power control means that the power will be continually changing. Once it has reached approximately the right level, then it will step up and then down by one level. In practice, the position of the mobile will change or the path will change as a result of other movements, and this will cause the signal level to move, so the continual change is not a problem.

Frames, slots and channels

UMTS uses CDMA techniques as its multiple access technology, but it additionally uses time-division techniques with a slot and frame structure to provide the full channel structure.

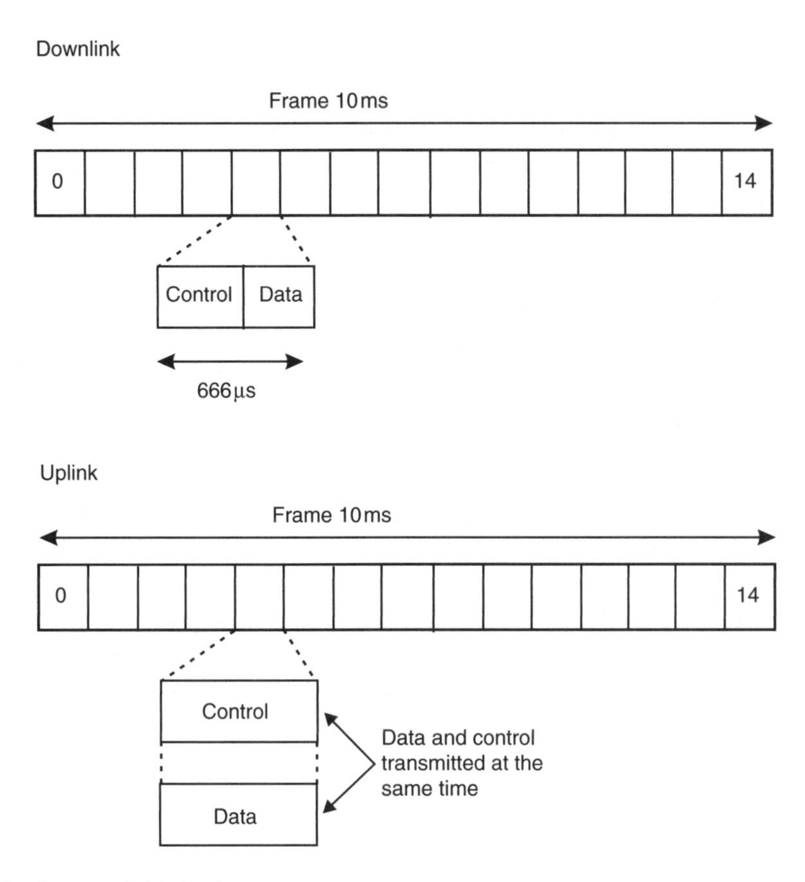

Figure 10.3 Frame and slot structure.

A channel is divided into 10-ms frames, each of which has fifteen time slots of 666 microseconds. On the downlink, the time is further subdivided so that the time slots contain fields that hold either user data or control messages (Figure 10.3).

On the uplink, dual channel modulation is used so that both data and control are transmitted simultaneously. Here, the control elements contain a pilot signal, Transport Format Combination Identifier (TFCI), FeedBack Information (FBI) and Transmission Power Control (TPC).

The channels carried are categorized into three groups: logical, transport and physical. The logical channels define the way in which the data will be transferred; the transport channels, along with the logical channel, also define the way in which the data are transferred; and the physical channels carry the payload data and govern the physical characteristics of the signal.

The channels are organized such that the logical channels are related to *what* is transported, whereas the physical layer transport channels deal with *how*, and with what characteristics. The MAC layer provides data transfer services on logical channels. A set of logical channel types is defined for different kinds of data transfer services.

Logical channels

1. *Broadcast Control Channel (BCCH)*, DL. The broadcast control channel is a downlink (DL) channel that broadcasts information relevant to that cell to the UEs in the service area. Information includes the radio channels of neighbouring cells with their configurations, as well as the available RACHs and scrambling codes.
2. *Paging Control Channel (PCCH)*, DL. This downlink channel is associated with the PICH, and is used for the transmission of paging messages as well as notification information.
3. *Dedicated Control Channel (DCCH)*, UL/DL. This channel is used to carry dedicated control information in both directions.
4. *Common Control Channel (CCCH)*, UL/DL. This bi-directional channel is used to transfer control information.
5. *Shared Channel Control Channel (SHCCH)*, bi-directional. This channel is bi-directional but is only found in the TDD form of UMTS, where it is used to transport shared channel control information.
6. *Dedicated Traffic Channel (DTCH)*, UL/DL. The DTCH is a bi-directional channel that is used to carry the user data or traffic.
7. *Common Traffic Channel (CTCH)*, DL. This unidirectional downlink channel is used to transfer dedicated user information to a group of UEs.

Transport channels

1. *Dedicated Transport Channel (DCH)*, UL/DL, mapped to DCCH and DTCH. This channel, which is present in both uplink and downlink directions, is used to transfer data specific to a particular UE. Each UE has its own DCH in each direction.
2. *Broadcast Channel (BCH)*, DL, mapped to BCCH. This is a downlink channel that transmits or broadcasts information to the UEs in the cell to enable them to identify the network and the cell. Typically, it provides information about system information, system configuration information and neighbouring radio channels. Of particular importance to UEs accessing the network, it identifies the available random access channels and their scrambling codes. The UE will also monitor the strength of the BCH to determine the relative signal strengths, and in this way it can determine which cell it should access. It is also used in determining whether a handover is required.
3. *Forward Access Channel (FACH)*, DL, mapped to BCCH, CCCH, CTCH, DCCH and DTCH. This channel appears in the downlink, and carries data or information to the UEs that are registered on the system. There may be more than one FACH per cell, as they may carry packet data. However, there is always one FACH that carries low-rate data to enable mobiles to receive the FACH messages, and it operates without closed loop power control.
4. *Paging Channel (PCH)*, DL, mapped to PCCH. Like the paging channels on other systems such as GSM, this channel carries messages that alert the UE to incoming calls, SMS messages, data sessions, or required maintenance such as re-registration.

5. *Random Access Channel (RACH)*, UL, mapped to CCCH, DCCH and DTCH. Again, this channel is similar in nature to the GSM random access channel, although this one carries requests for service from UEs trying to access the system. By its nature the RACH is a shared channel that any UE can access, and therefore mobiles requesting service will send a message. This is acknowledged by the network, using the Acquisition Indicator Channel (AICH). If the RACH request is not acknowledged, then the UE waits a random amount of time before sending another access request.
6. *Uplink Common Packet Channel (CPCH)*, UL, mapped to DCCH and DTCH. From its name it is understood that this is an uplink only channel. It provides additional capability beyond that of the RACH. If the mobile supports the use of this channel, it can access the system via this channel and then continue to use it for transmitting small amounts of data. This channel uses fast power control.
7. *Downlink Shared Channel (DSCH)*, DL, mapped to DCCH and DTCH. This channel can be shared by several users, and is used for data that are 'bursty' in nature – such as that obtained from web browsing, etc. It is time-shared by the different users, and the UMTS packet scheduling system dynamically assigns a particular slot on a carrier and code to a given UE that has packets to transmit in the required direction. Once the data have been sent, the packet scheduling system can assign the capacity to another UE.

Physical channels

1. *Primary Common Control Physical Channel (PCCPCH)*, mapped to BCH. This channel continuously transmits on the downlink, broadcasting the system identification and access control information. Its spreading code is permanently allocated to 256 so that it provides an overall transmission rate of 30 kbps. Half-rate convolutional coding is used, with interleaving over two consecutive frames.
2. *Secondary Common Control Physical Channel (SCCPCH)*, mapped to FACH, PCH. This is a downlink channel that carries the Forward Access Channel (FACH), providing control information, and the Paging Channel (PACH) with messages for UEs that are registered on the network. The spreading factor is determined according to the capabilities of the UE, allowing the fastest data transmission rate that the UE can handle.
3. *Physical Random Access Channel (PRACH)*, mapped to RACH. This is an uplink channel that enables the UE to transmit random access bursts in an attempt to access a network. In view of the nature of the channel a fixed spreading factor is used, giving a data rate of 16 kbps.
4. *Dedicated Physical Data Channel (DPDCH)*, mapped to DCH. This channel is used to transfer user data on both the uplink and downlink. The spreading factor depends upon the data rates, and also varies between the uplink and downlink because of the different modulation schemes used. On the downlink the spreading factor can vary between 4 and 256, and on the uplink it can vary from 4 to 512. Furthermore, the spreading factor can vary on a frame by frame basis (see Figures 10.4 and 10.5).
5. *Dedicated Physical Control Channel (DPCCH)*, mapped to DCH. This channel is present on both the uplink and downlink. It carries control information to and from the UE. The channel carries

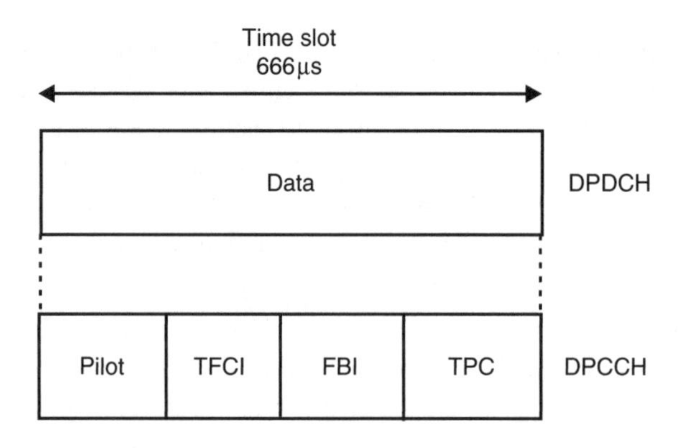

Figure 10.4 Uplink DPCCH and DPDCH slot structure.

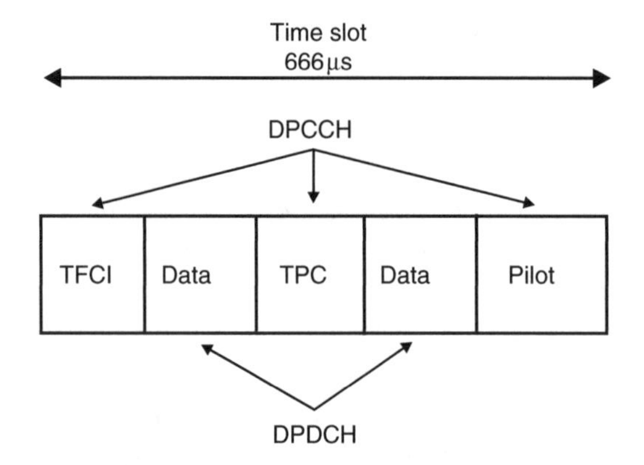

Figure 10.5 Downlink DPCCH and DPDCH slot structure.

pilot bits and the Transport Format Combination Identifier (TFCI) in both directions. The downlink channel also includes the Transmit Power Control and FeedBack Information (FBI) bits. Within the structure of the channel, the pilot bits aid the recovery of the control and data channel and the TFCI determines if multiple physical channels are combined (see Figures 10.4 and 10.5).

6. *Physical Downlink Shared Channel (PDSCH)*, mapped to DSCH. This channel shares control information to UEs within the coverage area of the node B. The channel is always associated with a downlink Dedicated Channel (DCH), and its spreading factor is able to vary between 4 and 256.

7. *Physical Common Packet Channel (PCPCH)*, mapped to CPCH. As might be gathered from its name, this channel is specifically intended to carry packet data. In operation, the UE monitors the system to check whether it is busy; if not, it transmits a brief access burst. This is retransmitted if no acknowledgement is gained, with a slight increase in power each time. Once the node B acknowledges the request, the data are transmitted on the channel.
8. *Synchronization Channel (SCH)*. The synchronization channel is used in allowing UEs to synchronize with the network as described above.
9. *Common Pilot Channel (CPICH)*. This channel is transmitted by every node B so that the UEs are able estimate the timing for signal demodulation. Additionally, it can be used as a beacon for the UE to determine the best cell with which to communicate. The CPICH uses a fixed spreading factor of 256 and this produces a data rate of 30 kbps.
10. *Acquisition Indicator Channel (AICH)*. The AICH is used to inform a UE about the Data Channel (DCH) it can use to communicate with the node B. This channel assignment occurs as a result of a successful random access service request from the UE.
11. *Paging Indication Channel (PICH)*. This channel provides the information to the UE so that it can operate its sleep mode to conserve its battery when listening on the paging channel. As the UE needs to know when to monitor the PCH, data are provided on the PICH to assign a UE a paging repetition ratio to enable it to determine how often it needs to 'wake up' and listen to the PCH.
12. *CPCH Status Indication Channel (CSICH)*. This channel, which only appears in the downlink, carries the status of the CPCH and may also be used to carry some intermittent (or 'bursty') data. It works in a similar fashion to the PICH.
13. *Collision Detection/Channel Assignment Indication Channel (CD/CA-ICH)*. This channel, present in the downlink, is used to indicate whether the channel assignment is active or inactive to the UE.

Packet data

Packet data provide an increasingly important element within mobile phone applications. W-CDMA is able to carry packet data transmissions in two ways. The first is for short data packets to be appended directly to a random access burst. This method is called common channel packet transmission, and is used for short infrequent packets. It is preferable to transmit short packets in this manner because the link maintenance needed for a dedicated channel would lead to an unacceptable overhead. Additionally, the delay in transferring the data to a dedicated channel is avoided.

Larger or more frequent packets are transmitted on a dedicated channel. A large, single packet is transmitted using a single-packet scheme, where the dedicated channel is released immediately after the packet has been transmitted. In a multipacket scheme, the dedicated channel is maintained by transmitting power control and synchronization information between subsequent packets.

Speech coding

Speech coding in UMTS uses a variety of source rates. As a result, a range of vocoders is employed, including the GSM EFR vocoder. When a variety of rates is available, a system known as Adaptive Multi-Rate (AMR) may be employed where rate is chosen according to the system capacity and requirements. This scheme is the same as that used on GSM and explained in Chapter 6. The actual vocoder that is chosen is governed by the system.

The speech coding process can be combined with a voice activity detector. This is particularly useful because during normal conversations there are long periods of inactivity. In the same way that discontinuous transmission is applied to GSM, it is applied to UMTS. It employs the same technique of inserting background noise when there is no speech, because when the discontinuous transmission cuts out the transmission no background noise would otherwise be heard, which can be very disconcerting for the listener.

Discontinuous reception

One of the big issues with mobile phones in general is that of battery life. It is one of the key differentiators that people take into account when buying a phone, and this gives a measure of its importance. Taking this into consideration when developing the UMTS/W-CDMA standard, a discontinuous reception or sleep mode was introduced. This mode allows several non-essential segments of the phone circuitry to power down during periods when paging messages will not be received.

To enable this facility to be introduced into the UMTS UE circuitry, the paging channel is divided into groups or sub-channels. The actual number of the paging sub-channel to be used by a particular UE is assigned by the network, and in this way the UE has to listen for only part of the time. To achieve this, the Paging Indicator Channel (PICH) is split into 10-ms frames, each of which comprises 300 bits – 288 for paging data and 12 idle bits. At the beginning of each paging channel frame there is a paging indicator (PI) that identifies the paging group being transmitted. By synchronizing with the paging channels being transmitted, it is able to turn the receiver on only when it needs to monitor the paging channel. As the receiver, with its RF circuitry, will consume power, savings can be made by switching it off.

Access stratum protocol layers

The MAC sub-layer is responsible for mapping the logical channels and the transport channels, ensuring optimum delivery of common and dedicated channel Protocol Data Units (PDUs). To deal with the range of channels, the MAC is subdivided into a variety of areas. These are the MAC-b, which is responsible for controlling the BCH; the MAC-c/sh, which controls access to the common transport channels except for the HS-DSCH; the MAC-d, which controls access to all dedicated transport channels; and the MAC-hs, which controls access to the HS-DSCH.

Sitting above the MAC layer is the Radio Link Control (RLC) layer, which is subdivided into three entities. These entities provide the control parts supporting three modes of operation at layer 2. The first is the Transparent Mode (TM), which transfers user data; and provides segmentation and re-assembly of data units. The second is the Unacknowledged Mode (UM), which is similar to the TM but also checks sequence numbers, and provides ciphering, padding and concatenation where required. Thirdly, there is the Acknowledged Mode (AM), which supports flow control, in-sequence delivery of upper-layer PDUs, error correction and duplicate detection.

The RLC also maintains the Quality of Service (QoS) as requested by the upper layers. This is required to ensure that a quality of service appropriate to the requirements of the application is received. There are four classes:

1. Conversational class (voice, video telephony, video gaming)
2. Streaming class (multimedia, video on demand, webcast)
3. Interactive class (web browsing, network gaming, database access)
4. Background class (email, SMS, downloading).

The top layer of the access stratum is called the Radio Resource Control (RRC) layer. This is a complex layer which has many functions. These include the control of broadcast information, the establishment and release of RRC connections and radio bearers, cell selection and re-selection, paging, ciphering, integrity protection, and the control of UE measurement reporting.

Using these features, the RRC layer provides the signalling connections used by the upper, non-access stratum layers, enabling information flow between the network and the UE.

There are two protocol states that can be adopted by the UE. The first is called Idle Mode, where the UE performs cell searches. The second is known as Connected Mode. Here, the behaviour of the UE is determined by the current state of the UE, of which there are four types: URA_PCH, CELL_PCH, CELL_FACH and CELL_DCH.

In the first two states, the UE is searching for the most appropriate cell and decoding broadcast system information. The UE can respond to paging in this mode. In the CELL_FACH mode, the UE can act on RRC messages received on the BCCH, CCCH and DCCH. In the CELL_DCH mode, the UE acts on messages received on the DCH.

Handover

Within UMTS, handover follows many concepts similar to those used for other CDMA systems such as CDMA2000. There are three basic types of handover: hard, soft and softer. All three types are used, but under different circumstances.

Hard handover is like that used for the previous generations of systems. Here, as the UE moves out of range of one node B the call has to be handed over to another frequency channel. In this instance, simultaneous reception of both channels is not possible.

Soft handover is a technique that was not available on the previous generations of mobile phone systems. With CDMA systems it is possible to have adjacent cell sites on the same frequency, and as a result the UE can receive the signals from two adjacent cells at once, while these cells are also able to receive the signals from the UE. When this occurs and handover is effected, it is known as soft handover.

The decisions about handover are generally handled by the RNC. It continually monitors information regarding the signals being received by both the UE and node B, and when a particular link has fallen below a given level and another better radio channel is available, it initiates a handover. As part of this monitoring process, the UE measures the Received Signal Code Power (RSCP) and Received Signal Strength Indicator (RSSI) and the information is then returned to the node B and hence to the RNC on the uplink control channel.

If a hard handover is required, then the RNC will instruct the UE to adopt a compressed mode, allowing short time intervals in which the UE is able to measure the channel quality of other radio channels.

Inter-system handover

While the primary handover strategy for a UMTS UE is to hand over to another UMTS node B, it may be the case that no other suitable UMTS node B is available, particularly during the roll-out phase of a 3G network. It is therefore expected that under these conditions the UE will perform an intersystem handover. As GSM and UMTS have completely different air interfaces, this form of handover is more complicated than a handover within UMTS or GSM. There are several reasons for this. Obviously, the 5-MHz channel and use of CDMA against the 200-kHz channel width and TDMA techniques are both hurdles to overcome. Another is the fact that UMTS continuously transmits data, and this does not allow the single-receiver UE to tune to another RF channel while communicating with the UMTS network.

To overcome this, a compressed mode has been provided within UMTS. This allows the UE to adapt its transmission rate to provide for interruptions in the transmission when it can monitor the GSM channels. These Transmission Gap Lengths (TGLs) are enabled by one of several methods – by lowering the amount of data to be transmitted, reducing the spreading factor or puncturing the error protection code. This feature is an essential element of UMTS, as it means that UEs can fall back to GSM (or GPRS/EDGE) when the full UMTS service is not available. Although it may appear to be particularly difficult, the fact that the updated GSM network has many common elements with the UMTS network, and may even use the same elements, does mean that the handover is simplified, at least a little.

The evolution of 3G networks

The maximum data transfer rate for UMTS is 2 Mbps. With Wi-Fi and other technologies, including wired ADSL links, providing much higher maximum download speeds and at lower cost, there was a need to ensure that UMTS did not fall behind and as a result lose business. To address this, high-speed data packet access technology was introduced into UMTS. Additionally, trends showed the volume of packet switched data rising and overtaking the more traditional circuit switched traffic. Using the new HSDPA scheme, it is possible to achieve peak data rates of 10 Mbps within the 5-MHz channel bandwidth offered under W-CDMA.

The first stage of the upgrade was to increase the data rate on the downlink channel, where there are typically more data transferred. This imbalance arises because of data downloads such as web browsing, video clips and the like. The resulting technology was known as High Speed Downlink Packet Access (HSDPA). In terms of the standards, Release 4 of the 3GPP W-CDMA standard provided the efficient IP support to enable provision of services through an all IP core network. Then Release 5 included HSDPA itself, with support for the packet-based multimedia services. In Release 6, antenna array processing technologies were introduced to enhance the peak data rate to about 30 Mbps. This involves smart antennas using beam forming for handsets with one antenna, and multiple-input multiple-output (MIMO) for UEs with up to four antennas.

As HSDPA needs to work alongside the original Release 99 systems, the new technology is completely backwards compatible.

One of the keys to the operation of HSDPA was the use of an additional form of modulation. Originally W-CDMA had used only QPSK as the modulation scheme; however, under the new system 16-QAM (which can carry a higher data rate but is less resilient to noise) was specified for when the link was sufficiently robust. The robustness of the channel and its suitability to use 16-QAM instead of QPSK is determined by analysing information fed back about a variety of parameters. These include details of the channel physical layer conditions, power control and quality of service, and information specific to HSDPA.

A scheme known as Hybrid Automatic Repeat Request (HARQ) was also implemented, along with multi-code operation, and this eliminated the need for a variable spreading factor. By using these approaches, all users, whether near or far from the base station, were able to receive the optimum available data rate.

Improvements in the area of scheduling were also made. By moving more intelligence into the node B, data traffic scheduling could be achieved in a more dynamic fashion. This enabled variations arising from fast fading to be accommodated, with the cell even being able to allocate much of the cell capacity for a short period of time to a particular user. In this way, the user was able to receive the data as fast as conditions allow.

To be able to implement the improvements, a new channel termed the High Speed Downlink Shared Channel (HS-DSCH) was introduced. W-CDMA normally carries data over dedicated transport channels (DCHs), several of which are multiplexed onto one RF carrier. This approach has been adopted because it provides the optimum performance with continuous user data. Under the new scheme, the 'bursty' nature of the data has been accounted for and more efficient use of the available spectrum has been made.

To enable all the HSDPA features to be implemented, a number of additional channels have been introduced. The High-Speed Shared Control Channel (HS-SCCH) is the downlink signalling channel that carries key physical layer control information. This information enables the demodulation of the data on HS-DSCH, and supports the data sent on HS-DSCH in the case of retransmission or an erroneous packet. The information includes the channelization code set, modulation scheme, transport block size, HARQ process information, redundancy and constellation version, and new data indicator.

The High Speed Dedicated Physical Control Channel (HS-DPCCH) is the signalling channel that carries the necessary control data in the up link. These consist of ARQ acknowledgements and the quality feedback information from the downlink, which are required by the node B. It also carries the downlink channel quality indicator (CQI) to indicate which estimated transport block size, modulation type, etc. can be received correctly with a reasonable error rate in the downlink.

The High Speed Physical Downlink Shared Channel (HS-PDSCH) is the physical channel that carries the user packet data in the downlink from the transport channel HS-DSCH. It has a fixed spreading factor of sixteen and multicode transmission using up to fifteen codes. The peak data rate is 10 Mbit/s, and it is modulated using 16-QAM.

HSDPA adapts to varying downlink radio and overall channel conditions by modifying the effective code rate, the modulation scheme, the number of codes used and the power per code. This enables the scheme to cope with a wide dynamic range and thereby give all users, regardless of their proximity or otherwise to the node B, the ability to make the best use of the high available data rates.

Fast packet scheduling functions are performed in the node B. These functions manage the HS-DSCH resources, and select the coding/modulation scheme and the power for the HS-DSCH data packets. The handsets that should be scheduled are determined using CQI reports coming from the UEs in question. Channel condition-dependent scheduling rather than sequential scheduling can increase the capacity significantly and make better use of air interface resources. Fast scheduling also allows guaranteed bit-rate services using packet scheduling, without the need for a dedicated channel. Handsets are prioritized by the scheduler according to the channel conditions, the amount of data pending in the buffer for each user, the elapsed time for that user since last served, and pending retransmissions for a handset.

HSDPA was the first to be implemented because most of the data are transported in the downlink direction. However, to support applications such as Voice over IP (VoIP) video-conferencing, and some of the other new ideas that are being introduced, high-speed data transmission is required in both directions. To this end, the equivalent for HSDPA, known as High Speed Uplink Packet Access (HSUPA), is also being introduced and is part of Release 6 of the 3GPP specification. This employs many of the same features found in HSDPA, but in the uplink direction. The upgrades include adaptive modulation along with HARQ, as well as improvements in the base station similar to those employed on HSDPA. While the maximum speed in the downlink direction is 14 Mbps, in the uplink it is less at 5.8 Mbps.

Position location

In recent years the importance of location services for mobile phone/cellular telecommunications has risen significantly. One of the major reasons for this is the requirement to locate mobile phones in the case of an emergency. For landlines it is a simple matter to trace the location of the call, because by their very nature they are fixed. For mobile phones the matter is quite different, and on many occasions the emergency services have had great difficulty in locating mobiles with any degree of accuracy. With rulings such as the E911 directive in the USA requiring operators to be able to locate mobile phones accurately, the concept of phone position location has become more important.

A number of techniques are made available by cell phone operators, including cell identity (Cell ID), Time Difference of Arrival (TDOA), Enhanced Observed Time Difference (EOTD), Advanced Forward Link Trilateration (AFLT) and Assisted GPS (A-GPS). Of these, Cell ID, TDOA and A-GPS are being widely used, with A-GPS emerging as the front runner. The other schemes have not gained any real acceptance and commercial usage.

Cell ID

Cell identity is the most basic positioning system used in GSM, and for many years was the only method of locating a mobile. It is a network-based method, and utilizes the fact that once the mobile has registered with the network, the network has to know where the mobile is so that it can be paged when there are incoming calls or text messages, etc. This information is available from the network and the handset.

With the cell identified, the cell ID information is converted to a geographic position using parameters from the network. This information is available in the coverage database at the Serving Mobile Location Centre (SMLC).

The major drawback of this system is the level of accuracy. The location of the mobile can only be determined to within a given cell. This will depend on the size of the cell, which in turn depends on a number of factors. Cells for networks using the lower-frequency bands, such as 900 MHz, tend to be larger than those for networks operating at 1800 MHz. The terrain will have a large effect, as will the level of usage. In open country where usage levels are low, macrocells might be installed that may have a radius of up to 35 km, whereas urban cells, where usage is much higher and signals will not propagate as far, tend to be much smaller and may have a radius of 1 km or less. Some picocells may only have a very small coverage area, and it would be possible to locate a mobile accurately if one of these was used.

Moreover, many cells are sectored, and this will further narrow the area in which the mobile is located. A sectored cell uses a number of antennas to split the cell; three are often used, providing a coverage of 120°.

While Cell ID is limited in its accuracy, it has the advantage that it is possible to locate any cell phone, however old, once it is registered onto the network.

TDOA

Time Difference of Arrival (TDoA) is another scheme that is in use, particularly in the USA. It provides better location than cell ID, and operates by measuring the relative time of arrival of signals transmitted by three base stations at the handset, or the signal transmitted by a mobile at three base stations. As the radio signals travel at the speed of light, it is very easy to relate the time of travel to a distance from the base station. As no absolute measurement is possible, the time differences between signals are measured. The time difference between two signals enables a hyperbolic line to be traced out on a map along which the mobile may be located. When using the time differences between three stations, the hyperbolae intersect, providing a unique location.

In order for this technique to work, there must be very accurate synchronization between the base stations. Any differences will manifest themselves as offsets from the true position. However, unlike some techniques, it does not require a large amount of data to be transmitted over the network for it to operate. In view of the fact that data capacity is limited, any method of reducing it is welcomed by network operators.

A-GPS

The most accurate technique, and that which seems to be gaining the widest level of acceptance, is known as assisted GPS. It uses the Global Positioning System (GPS) as the basis of its operation, but takes additional data to enable the GPS to operate under weak signal conditions, and to provide the service required for mobile phone location.

Prior to looking at how A-GPS operates, a brief description of the operation of GPS is provided so that the requirement for assistance can be understood.

GPS

The full operational system consists of a constellation of twenty-four operational satellites, with a few more in orbit as spares in case one fails. The satellites are in one of six orbits, which are in planes that are inclined at approximately 55° to the equatorial plane. There are four satellites in each orbit. This arrangement provides the user with a view of between five and eight satellites at any time from any point on the Earth. When four satellites are visible, sufficient information is available to be able to calculate the user's position on Earth.

The GPS satellites need to be monitored and controlled from the ground, and it is necessary to be in contact with each satellite for most of the time to be able to maintain the level of performance required. To achieve this, there is a master station located at Falcon Air Force Base, Colorado Springs, USA. However, other remote stations are located on Hawaii, Ascension Island and Diego Garcia, and at Kwajalein. Using all these stations, the satellites can be tracked and monitored for 92 per cent of the time. This results in two 1.5-hour periods every day when each satellite is out of contact with the ground stations. This results in two 1.5 hour periods every day when each satellite is out of contact with the ground stations.

Figure 11.1 shows a GPS satellite being launched into orbit.

Using the network of ground stations, the performance of the GPS satellites is monitored very closely. The information that is received at the remote stations is passed to the main operational centre at Colorado Springs, and the received information is assessed. Parameters such as the orbit and clock performance are monitored, and action taken to reposition the satellite if it is drifting even very slightly out of its orbit, or to adjust the clock if necessary or, more usually, to provide data to it indicating its error. This information is passed to three uplink stations co-located with the downlink monitoring stations at Ascension Island, Diego Garcia and Kwejalein.

The GPS operates by a process of triangulation. Each GPS satellite transmits information about the time, and its position. By comparing the signals received from four satellites, the receiver is able to deduce how long it has taken for the signals to arrive. From knowledge of the position of the satellites, it can calculate its own position.

The GPS satellites transmit two signals on different frequencies. One is at 1575.42 MHz and the other is at 1227.6 MHz. These provide two services, one known as course acquisition (C/A) and the other as a precision (P) signal. The precision signal is only available for the military, but the C/A elements of GPS are open to commercial use.

Both signals are transmitted using direct sequence spread spectrum (DSSS), and this enables all the satellites to use the same frequency. They can be separated in the GPS receiver by the fact that they use different orthogonal spreading codes, and this works in exactly the same way as the CDMA cell phone systems. The spreading codes are accurately aligned to GPS time to enable decoding of the signals to be facilitated.

The coarse acquisition signal at 1.5 GHz uses a 1.023-MHz spreading or chip code, while the precision signal is transmitted at 1.2 GHz using a 10.23-MHz code. This precision signal is encrypted, and uses a higher power level. Not only does this assist in providing a higher level of accuracy; it also improves the reception in buildings.

Figure 11.1 A GPS satellite being launched into orbit (image courtesy of NASA).

All the GPS satellites continually transmit information, including what are termed Ephemeris data, almanac data, satellite health information and clock correction data. Correction parameters for the ionosphere and troposphere are also transmitted, as these have a small but significant effect on signals even at these frequencies.

The Ephemeris data provide information that enables the precise orbit of the GPS satellite to be calculated. The almanac data give the approximate position of all the satellites in the constellation, and from this the GPS receiver is able to discover which satellites are in view. Although each satellite contains an atomic clock, they all drift to a small extent and as a result details of the clock offsets are transmitted. It is found that it is more effective to measure the error and transmit the data than to maintain the clock exactly on time.

The data transmitted by the GPS satellite are formatted into twenty-five frames, each 1500 bits in length. The frames are divided equally into five sub-frames. At a data transmission rate of 50 bits per second, it takes 6 seconds to transmit a sub-frame, 30 seconds to transmit a frame and 12.5 minutes to transmit the complete set of twenty-five frames.

Sub-frames 1, 2 and 3 are the same for all data frames, and contain critical satellite-specific information. This allows the receiver to determine a single satellite clock correction and Ephemeris within 30 seconds. Sub-frames 4 and 5 contain less critical data that apply to the complete satellite constellation, and these data are distributed throughout the twenty-five frames.

The strength of the GPS signal that is received on the surface of the earth is very low. Typically it is around −127 dBm, although there is a variation on this arising from the elevation of the GPS satellite. This may reduce the signal level by up to about 3 dB. To receive the signal a receiver bandwidth of around 2 MHz is often used, even though a chip rate of 1.023 MHz is employed. The reason for the wide bandwidth is that it reduces differential group delay, which would cause positional errors.

With the wide bandwidth and low signal levels, the actual received signal is below the thermal noise level. The only way to recover the signal is by correlation over a large number of chips. Commonly, the correlation is done over a complete code cycle of 1023 chips, giving a correlation time of 1 ms. Using these techniques, the best receivers may receive signals down to levels of around −142 dBm.

From a 'cold start', the GPS receiver chooses a satellite to look for and tries all possible code phases to see if correlation is achieved. The problem is compounded by the Doppler shift on the satellite, which forces the receiver to look in a number of Doppler 'bins' for each code phase, thus increasing the search time. If no satellite is found, then the search is repeated for the next satellite. Modern receivers speed up this search by using large numbers of correlators in parallel. Tens of thousands of correlators are typically used.

Once the first signal has been correlated, the GPS receiver can demodulate the data the signal carries. With the almanac data available, the GPS receiver is able to deduce which satellites are visible and hence which ones to receive. Additionally, this enables the receiver to correlate the signals more quickly.

The receiver measures the relative phases of the signals from each of the satellites to provide what are termed 'pseudoranges'. The GPS receiver then uses the Ephemeris data and also compensates for elements such as the clock offsets, effects of the ionosphere and troposphere, and even relativity. The receiver uses all of this information to calculate its own clock error and position. The overall calculation is somewhat involved, and uses iterative processes to reach the final result.

In view of the time taken to correlate with the GPS satellites, as well as the time taken to transmit the data, the Time To First Fix (TTFF) is usually in excess of 12.5 minutes. Faster TTFF times from a cold start are often achieved by using a vast number of correlators, and this approach means that it is not always necessary to wait until all the data have been received before the first fixes can be made.

Assistance from the base station

The problem with GPS for mobile phone applications is that signal levels are low, and the receiver needs to have a direct view of the satellite. This can cause problems when the phone is used in a building, or even in an urban area where a direct view of the satellite is masked. Additionally, the time taken for the receiver to lock – the Time To First Fix (TTFF) – can be as much as 10 minutes or more from switch-on. This is not acceptable when emergency calls are being made, as a much faster acquisition time is required. To achieve this, assistance is required for the GPS receiver, thereby giving rise to the requirement for Assisted GPS.

A-GPS

The system known as Assisted GPS or A-GPS uses the mobile phone network to assist the GPS receiver in the mobile phone to overcome the problems associated with TTFF, and the low signal levels that are encountered in some situations.

For A-GPS, the network provides the Ephemeris data to the cell phone GPS receiver, and this improves the TTFF. This can be achieved by incorporating a GPS receiver into the base station itself, and as this is sufficiently close in position to the mobile, the data received by the base station are sufficiently accurate to be transmitted on to the mobiles. The base station receiver is obviously on all the time, and will be located in a position where it can 'see' the satellites.

The information provided can be either the Ephemeris data for visible satellites or, more helpfully, the code phase and Doppler ranges over which the mobile has to search – i.e. 'acquisition data'. These ranges can be estimated as the position of the mobile is bounded, because it must be within the cell served by the particular base station. This technique is able to improve the TTFF by many orders of magnitude.

Assisted GPS is also used to improve the performance within buildings where the GPS signals are attenuated by 20 dB or possibly more. Again, by providing information to the GPS receiver in the mobile it is able better to correlate the signal being received from the satellite when the signal is low in strength. Using this technique it is possible to gain considerable increases in sensitivity, and some manufacturers have claimed it is possible to receive signals down to power levels of around −159 dBm. The base station supplies the receiver with navigation message bits – 'sensitivity data'.

Several techniques are used to pass the assistance data from the base station to the mobile. For GSM, GPRS, EDGE and CDMA2000, the information is transmitted as packet data requiring a new session to be activated. For UMTS, a data channel is used and the information is broadcast to the users. In this way, the level of data is reduced.

Conformance and interoperability testing

Testing of mobile phone designs before they enter service and are used on a live network is particularly important. Users expect their phones to operate correctly – to be able to make calls, and for all the features to work as intended – under all conditions. The huge variety of phones on the market means that a user can easily change to another manufacturer, and in a market that is very sensitive a manufacturer can soon get a poor name. With the vast number of phones of any given design being produced, recalling them is not normally a viable option. Additionally, it has been known for incorrectly operating phones to cause problems to the network operator.

Ensuring that phones operate virtually perfectly is not an easy task. The specifications used to define the standard are usually very extensive, and can be open to interpretation in some places. Essential operations such as registration require many transactions between the phone and the network, and as a call proceeds the network and phone have to keep in communication with one another to ensure that the call is maintained. Handover and other scenarios then introduce further complications. All these interactions must proceed correctly, despite the huge number of communications that are undertaken under a wide variety of scenarios. With the introduction of 3G these protocol exchanges have become increasingly complicated, and still demand the same high level of operation. In addition to this, many physical elements of the phone need to be tested – from ensuring that the phone can be heard, to checking that people using them do not absorb too much RF energy.

Types of test

To ensure that a phone meets its required standards, it has to undergo a variety of types of test. It should naturally pass basic safety tests to ensure that there is no way that it can cause harm or injury.

Similarly, it also needs to undergo what is called Specific Absorption Rate (SAR) testing. This involves the use of an anatomically correct model of the human head. Sensors are set up inside the model to measure the temperature rises to ensure that the heating effect is within acceptable limits.

It is necessary to check the protocol operation of the phone and, with the complicated protocols used in mobile phones, this is a critical area. If the phone protocol software operates incorrectly, then it could result in problems experienced not only by the phone but also on the network. In view of the complexity of the protocols, this testing can be very involved. Specialized network simulators are used, and these testers emulate a variety of network entities – i.e. base stations or node Bs, RNC (Radio Network Controller) and the like. In this way a host of scenarios from registration to handover, and in fact any situation that can be encountered, can be simulated.

RF testing is also undertaken. Many measurements of the transmitter and receiver performance are undertaken in a variety of areas, such as the out-of-band emissions. Measurements of the Radio Resource Management (RRM) are undertaken to ensure that the control capability of the phone is operating correctly. There are, for instance, very tight limits on the control of the transmitter output power to ensure that the phone radiates only as much as is needed under any given condition, and that the noise in the phone bands is reduced to the minimum level. To achieve this, a protocol tester is often used to control the phone and set up the relevant scenarios. In addition to this, RF measurement and generation equipment is required. This is often in the form of additional signal generators, power meters, analysers, noise generators, etc. To check operation of the phone with multi-path and fading, special fading simulators are required.

It is also necessary to test the operation of the SIM or, in the case of 3G UMTS phones, the USIM. As SIMs are interchangeable between phones, the interface must be rigorously checked. It is also vital to verify the security aspects of the operation of the SIM, as lapses in security could compromise elements of the network security. To undertake this testing, a SIM simulator (or USIM simulator) is required. This simulator emulates the operation of the SIM, and tests on the phone can then be run using a protocol tester to set up the variety of scenarios that are needed.

Finally, audio checks are undertaken. These check the correct operation of the audio aspects of the phone, in terms of both the microphone and the earphone. Checks of audio levels, quality and much more are measured, using a variety of equipment, to ensure they conform to the requirements laid down.

Test cases

It is obviously necessary to ensure that each stage in the testing is repeatable, regardless of the test equipment used and the organization performing the testing.

To achieve uniformity, a large number of what are termed 'test cases' are defined. These were originally defined by ETSI (European Telecommunications Standards Institute) for GSM, but they are now controlled by 3GPP (Third Generation Partnership Project).

For GSM these test cases were written in prose, and they described the test itself, the set-up conditions and the applied stimuli, and of course the pass and fail criteria. It was then possible for each test equipment manufacturer to implement these tests on their equipment. However, to ensure that the tests are a faithful implementation of the original intent, a validation and certification process has been set up.

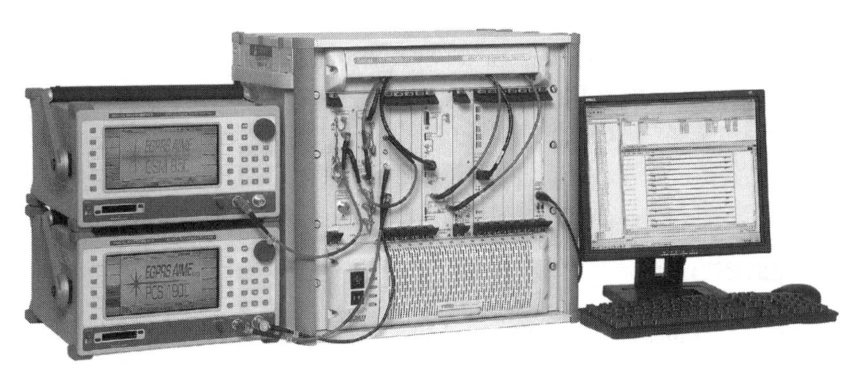

Figure 12.1 A system used for conformance testing the handover between 3G and 2/2.5G (image courtesy of Aeroflex).

Once the manufacturer is content that the test operates satisfactorily, the test is forwarded to an independent validation organization that will assess it for conformance with the original test case (see Figure 12.1). Assuming that the test case satisfactorily passes this process, then it is presented to an industry body for certification. Once it has achieved this status, it can then be used for formal handset testing and certification.

The organization that has overall control of the test cases for GSM and UMTS is 3GPP. However, changes are handled by the GERAN (GSM Enhanced Radio Access Network) working Group for GSM, and a group known as the T1 group for UMTS. The validation and approval of the implemented test cases is then handled by the GCF (Global Certification Forum).

The North American version of GSM running in the 1900-MHz band is often referred to as PCS (Personal Communications System), and there is another allocation at 850 MHz. A group known as the PVG (PCS Validation Group) handles the approvals, and its results are ratified by the PTCRB (PCS Type Certification Review Board). Phones are then tested against the test cases and, if successful, are certified by the CTIA. To achieve CTIA certification, it is necessary for phones to be tested in CTIA-approved laboratories.

As might be expected, experience gained on GSM has been reflected in improvements for the new third-generation UMTS system. One of the main changes is that the protocol test cases are initially written in prose and then converted into TTCN (Tree and Tabular Combined Notation) to be made available to the industry. This language enables the test cases to be compiled into a format that can be run directly on the target test equipment. This approach saves time for the industry as a whole and also reduces costs, because generating the test cases is far easier to achieve. The main advantage is that it gives far more consistency across the industry, as tests are no longer open to the same level of interpretation that they were before. This saves time in the validation process.

The TTCN test cases have been prepared by a team of industry experts based at ETSI and working on behalf of 3GPP. They prepared the basic TTCN code that was reviewed within the 3GPP community using email reflectors. By reviewing the software in this way at the beginning of the process, the individual test cases do not need to be reviewed each time they are submitted by a test equipment manufacturer for validation.

To test a new handset, the manufacturer generally approaches a qualified test house which will possess a variety of different test systems. The test house will then run the required certified test cases and be able to present a case for a handset being suitable for use on the available networks.

CDMA system

A different approach to that used by GSM/UMTS has been adopted by the CDMA-based systems. The emphasis of testing is on interoperability with the network, rather than conformance with a specification. This has the advantage that it checks that the phone actually works on the network.

For CDMA2000, the standards are written by 3GPP2. This is the equivalent of 3GPP, but for the CDMA standard IS-2000. These standards are then published by the relevant standards body for a particular area. In North America, this is the TIA (Telecommunications Industry Association); for Japan it is the ARIB (Association of Radio Industries and Businesses), etc. In North America, testing is carried out under the auspices of either the CDG (CDMA Development Group) or the CTIA.

Testing is then conducted in three stages. Stage 1 testing verifies the RF performance of a system, checking parameters such as receiver sensitivity, and performance under fading conditions. In addition to this, protocol testing is undertaken. Stage 2 is what is termed 'cabled interoperability testing'. Here, the focus is on the interoperability of a handset with a particular base station. If a network has base stations from one manufacturer, then the handset will be tested against this. In this way, any operational anomalies between the two will be discovered. Tests ensure that the mobile performs in the correct manner with the base station under a variety of conditions, including call set-up, handoff and call termination. Finally, Stage 3 testing is undertaken by an operator and consists of running the mobile on a live network to ensure that it works over the air. This may include what is termed a 'drive test', where the mobile is driven around a live network and its performance checked under real conditions.

The testing carried out under CTIA auspices is undertaken by approved laboratories. These use test equipment to simulate the network, and tests similar to those in CDG Stages 1 and 2 are employed. Once mobile phones have successfully passed the CTIA tests, they are given an approval certificate.

Different test requirements are placed upon handset manufacturers by the various network operators, and therefore phones may undergo a variety of tests under either CDG or CTIA auspices, or both.

Although the idea of interoperability testing is well established, there is a move towards the conformance testing approach. Part of the reason behind this is to ensure a higher degree of commonality between phones. This will facilitate aspects such as roaming, where there is a greater need for phones to work on a variety of networks. This aspect has been seen as one of the major successes of GSM, as a GSM phone will work on any network around the globe. Currently CDMA phones tend to be targeted towards a particular network, and there may be minor differences between networks.

Summary

With well over a billion and a half subscribers connected to GSM networks alone, and many millions of phones being produced each year, testing is a major element in the production of a mobile phone design. It is particularly important to operators that phones work correctly. If they do not operate correctly, it is likely that users will change their network as well as the phone, not knowing where the problem lies. With the enormous levels of competition between the variety of network operators and also between the different phone manufacturers, none can afford poor levels of performance. Adequate testing in the development stages can save considerable amounts of money later – far outweighing the cost of any test equipment purchased.

Glossary

1X RTT The abbreviation used to describe the first phase of CDMA2000 that uses a single 1.25-MHz bandwidth RF carrier channel. The abbreviation RTT stands for Radio Transmission Technology spreading, and the 1X element indicates that it uses the spreading required for one 1.25-MHz channel.

3GPP Third Generation Partnership Programme. The standards body and organization used to develop and coordinate W-CDMA/UMTS development.

3GPP2 Third Generation Partnership Programme 2. The standards body and organization used to coordinate the development of CDMA2000 in the USA.

3X RTT The version of CDMA2000 that will be capable of higher data rates than 1X systems. It uses a spreading rate of three times the basic 1X technology. RTT stands for Radio Transmission Technology.

ACELP Algebraic Code Excited Linear Prediction. A vocoder technique used in GSM and NA-TDMA.

ADPCM Adaptive Differential Pulse Code Modulation. A form of speech coding widely used for wireless local loop networks and cordless phones employing a 32-kbps data rate.

Air interface The interface between the mobile phone and the base station. Often this refers to the protocols used in this area.

AM Amplitude Modulation. A form of modulation used widely on the long-, medium- and short-wave broadcast bands. It involves changing the amplitude of the signal in line with the instantaneous amplitude of the incoming audio or modulating signal.

AMPS Advanced Mobile Phone System. The analogue first-generation mobile phone system used extensively in North America and many other countries.

ANSI American National Standards Institute. The main North American standards approving body.

ARQ Automatic Repeat reQuest. A form of error correction system whereby when errors are detected by the receiver, it sends a repeat request so that the data with errors are re-sent.

AuC Authentication Centre. The database used to control the authentication process and compare the users' identification with those recognized as valid by the network.

Authentication A process used to validate a user before allowing the user onto the network.

Authentication key The secret 64-bit word used in the North American authentication process.

BER Bit Error Rate. The rate at which errors appear in a digital transmission system. A measure of the bit errors is often used to determine factors such as the receiver performance or link quality.

BSC Base Station Controller. The part of the network that controls one or more base stations and interfaces with the switching centre (e.g. the MSC used in GSM).

BSS Base Station System. The overall system that encompasses the BTS and BSC.

BTS Base station transceiver system. The base station equipment used to transmit and receive signals to and from the mobile handsets.

Burst This term is widely used in TDMA, and refers to one time slot or transmission (in multi-slot operation, one burst may include several time slots used by the same user for high-speed data transmission). In CDMA systems, it refers to one power control group.

CAMEL Customized Applications for Mobile Enhanced Logic. This refers to the enhanced intelligent network capabilities being added to GSM.

CDMA Code Division Multiple Access. A system based around direct sequence spread spectrum that enables multiple users to access a cellular telecommunications network simultaneously. By using different spreading codes several 'channels' can be created, and these can be allocated to different users and facilities on the same physical RF channel.

Cellular radio A radio system that uses a number of small 'cells' within which the signal is transmitted to a number of users. The user can then move from one cell to the next while still maintaining a call. In view of the relatively small size of the cells, frequencies can be re-used at some distance, enabling efficient occupancy of the spectrum.

Circuit switching The process where a dedicated channel or circuit is allocated to a given user until a call is completed. This has the advantage of being a (relatively) simple approach, and does not require as much management as other approaches such as packet switching.

Common control channel A channel that carries information for all the mobile handsets communicating with a base station.

Constellation analysis A graphical method of analysis of a digitally modulated carrier. It is used primarily for phase-modulated signals.

CSS Cellular Subscriber Station. Another name for a mobile handset or terminal.

CTIA Cellular Telecommunications Industry Association. The North American organization that includes wireless manufacturers and carriers.

DAMPS Digital AMPS. The digital system that followed on from AMPS. It is more commonly known as NA-TDMA.

DCS Digital Cellular System. This is a version of GSM operating at 1800 MHz. It was a term used shortly after the introduction of GSM operating at 1800 MHz. Nowadays, it is another band for GSM, sometimes called GSM 1800. There is a further band at 1900 MHz, and this too may occasionally be referred to as

	DCS 1900 although the original term was PCS 1900 (Personal Communications System).
DECT	Digital Enhanced Cordless Telephone. A European cordless telephone standard.
Diversity	A receiving technique employing two antennas spaced apart from each other so that reception conditions will be different for both. The receiver then switches between antennas to ensure the best reception is maintained at all times.
Downlink	This is also known as the forward link, and is the transmission path from the base station to the mobile.
DSSS	Direct Sequence Spread Spectrum. A form of transmission used for a number of applications, including CDMA phone technology, where data are multiplied by a spreading code to spread the signal over a wide bandwidth. Only when it is correlated with the same spreading code in the receiver can the data be extracted.
DTx	Discontinuous Transmission. This is a method of reducing the transmission time of a mobile handset by cutting the transmission when no audio is present.
Dual band	Phones that operate on two bands, but with one standard or radio access technology. Many phones may be triple or even quad band now.
Dual mode	This refers to phones that are able to operate on two standards or radio access technologies.
Dual Tone Multiple Frequency	A method of transmitting numbers over a phone line by using two tones, different combinations of which represent different numbers.
E_b/N_0	This term is the energy per bit divided by the noise power spectral density. The energy per bit is simply the signal power, and hence this term is effectively a form of signal-to-noise ratio.
EDGE	Enhanced Data for Global Evolution. Sometimes referred to as Enhanced Data for GSM Evolution, this is a system that uses packet switched data along with 8PSK modulation and other updates to provide enhanced data rate transfer over what is basically a GSM network.
EIR	Equipment Identity Register. A database used within GSM that retains information about whether a mobile phone has been stolen.
Enhanced Variable Rate Vocoder	An 8-kHz vocoder that can be used on CDMA networks to provide improved voice quality of the QCELP vocoder normally used.
Erlang	A measure of radio channel usage – 1 Erlang is equivalent to one telephone line being used permanently.
ERP	Effective Radiated Power. The level of power radiated by an antenna in the required direction.
FDD	Frequency Division Duplex. The technique where transmission and reception (i.e. forward and reverse channels) use different frequencies.
FDMA	Frequency Division Multiple Access. The multiple access scheme that allows different users access to a cellular telecommunications network by assigning them different frequencies.

FM	Frequency Modulation. A system where the frequency of a radio frequency carrier is changed in line with the audio or data to be transmitted. It is this type of modulation that is used by many walkie-talkie transceivers, and for FM broadcasting.
Forward link	This is also known as the downlink, and is the transmission path from the base station to the mobile.
Frequency hopping	A process where the frequency of a transmission is changed periodically, hopping from one frequency to another, normally in a pseudorandom fashion. It has the effect of spreading any interference more evenly between mobiles, and as such it is less noticeable.
GMSK	Gaussian Minimum Frequency Shift Keying. A form of frequency shift keying that keys on a zero crossing point to reduce bandwidth, and is then further filtered with a Gaussian filter to provide the required bandwidth.
GPRS	General Packet Radio Service. An update to GSM whereby packet switching rather than circuit switching is used to provide increased data throughput.
GSM	Global System for Mobile telecommunications. Originally named Groupe Speciale Mobile, it was renamed when its use spread outside Europe. It is a digital TDMA system based on 200-kHz RF channel bandwidths.
Half-rate	This term refers to a vocoder that is able to use half the bandwidth of existing or established vocoders, and in this way reduces the amount of bandwidth required and increases the number of calls that can be handled.
Handoff	This is the North American term for the process of transferring a mobile handset from one base station or base station sector to another.
Handover	This is the European term for the process of transferring a mobile handset from one base station to another.
HLR	Home Location Register. A database within the network that contains information about the mobiles that subscribe to that particular network.
HSCSD	High Speed Circuit Switched Data. A service that allowed data at speeds up to 64 kbps to be transmitted over a GSM network. It was superseded by GPRS.
HSDPA	High Speed Packet Downlink Access. An update to the original UMTS standard that allows it to carry data on the downlink at much higher rates than were possible in the original releases of the standard.
HSUPA	High Speed Packet Uplink Access. An update to the original UMTS standard that allows it to carry data on the uplink at much higher rates than were possible in the original releases of the standard.
IMEI	International Mobile Equipment Identity. A serial number electronically installed in a mobile handset to identify it. This can be used to immobilize or track stolen handsets.
IMSI	International Mobile Subscriber Identity. A serial number (normally contained within a SIM) that identifies the subscriber, and hence the telephone number.

InterRAT handover A handover that takes place between two different radio access technologies. This occurs when the coverage of one system does not extend as far as that of another. This occurrence is widespread whilst 3G is being deployed and handsets need to hand over from a 3G service to a 2G service as the handset moves outside the coverage area of the 3G network. It is less of an issue with CDMA2000 services, as there is backward compatibility between the different levels of service.

Intersymbol interference A form of interference caused by reflections in the signal path giving rise to some portions of the signal travelling over different path lengths. This results in time delays, which can mean that different symbols arrive at the same time, resulting in interference between them.

Intersystem handover (handoff) A handover that takes place when a mobile hands over from one radio access technology to another.

Intracell handover (handoff) A handover from one frequency to another within the same cell.

LPC-RPE Linear Prediction Coding with Regular Pulse Excitation. A vocoder technique used in GSM.

Macrocell A cell covering a large area, often several kilometres in diameter.

MAHO Mobile Assisted Handover (Handoff). This occurs when the mobile handset assists in the handover process by making measurements, for example of signal strength, to assist the base station in the handover process.

Microcell A cell covering a very small area, often less than 500 metres in diameter. Typically one is created by having a low-power base station and placing the antenna below the height of the surrounding buildings. In this way the radiation will follow the pattern of the street.

Modulation The process of changing the radio frequency carrier signal in line with the audio signal or data to be transmitted. This modulation is then extracted by the receiver, and the original audio or data can be recreated.

MSC Mobile Switching Station. The switching centre where the mobile network interfaces to the public telephone system.

MTSO Mobile Telephone Switching Office. The switching centre used in the AMPS system that controls base stations and also interfaces to the public telephone system.

NAMPS Narrowband AMPS. A narrowband version of the AMPS standard developed by Motorola using a 10-kHz channel bandwidth rather than the 30 kHz used by AMPS. It was aimed at overcoming the problem of spectrum crowding. As cellular technology became more widespread, voice quality was poorer, and it was overtaken by 2G technology. In view of these reasons, it was not widely deployed.

NA-TDMA North American TDMA. The digital second-generation system first adopted in the USA. It uses time division multiple access (TDMA) technology, and this has given rise to its name. Also known as Digital AMPS and US Digital

	Cellular, it is specified under standard IS-136, having been initially issued as IS-54.
NMT	Nordic Mobile Telephone. The analogue (1G) standard that was developed in Scandinavia and was the first mobile phone system to be commercially deployed (1979).
Node B	A terminology used within UMTS to denote a base station.
Open loop	This refers to the fact that CDMA systems in particular require considerable degrees of power control. Such control is normally undertaken with feedback loop in operation between the base station and the mobile handset. However, there are some instances (for example when a handset has just been switched on and is registering) when this information is not available. The handset will then operate 'open loop', making an estimate of the power required, dependent upon the signal it is receiving from the base station.
Origination	The process of starting or originating a call. This is done by the handset when the user dials a call.
Orthogonal codes	The codes used as spreading codes when generating DSSS (CDMA) signals. They are said to be orthogonal if, when they are summed over a period of time, they add to zero.
Packet switching	The process whereby data are segmented into packets, each of which is typically a few milliseconds long. These packets are then transmitted along whatever channel is available. As there are normally dead periods when no data are transmitted by a particular user, this system makes more efficient use of the available capacity than does the circuit switched approach.
Paging channel	The channel used by the mobile handset in GSM and CDMA to monitor whether there is an incoming call about to arrive.
PCS	Personal Communications System. This normally refers to GSM used on the 1900-MHz band. It is now just called GSM or, occasionally, GSM 1900.
PDC	Pacific Digital Cellular, sometimes also referred to as Personal Digital Cellular. The 2G TDMA system similar to NA-TDMA that is widely used in Japan.
PLMN	Public Land Mobile Network. The landline or wired side of the mobile network, i.e. the infrastructure.
PM	Phase Modulation. A form of modulation where the phase of the signal is modulated in line with the audio or data to be transmitted. This form of modulation is used in many data applications.
PN	Pseudorandom number. A number that is generated mathematically to create an almost random number sequence. By using the same start or 'seed' number, the number can be recreated and therefore it is not fully random.
PSTN	Public Switched Telephone Network. The public landline-based telephone network.
QCELP	Qualcomm Code Excited Linear Prediction. The vocoder used in cdmaOne/CDMA2000.

Rake receiver A receiver used for reception of CDMA signals that enables multi-path signals to be coherently combined to reinforce one another.

RAT Radio Access Technology. This refers to the type of cellular system that is in use. GSM, NA-TDMA, W-CDMA and CDMA 2000 are all different radio access technologies.

Registration The process whereby a mobile gives information to the network so that it can gain access.

Reverse link A term used to describe the link from the mobile to the base station.

RF Radio Frequency. Pertaining to those parts of a system that manipulate the signals at high or radio frequencies.

Rho This is a transmitter performance measurement that gives an indication of transmitter quality.

Roaming The process that enables a mobile to be used on a different network to the one for which it was originally intended. It occurs when travelling outside the coverage area of the original network, e.g. when abroad.

SID System Identifier. A code sent by base stations to indicate a particular operator or geographic area.

SIM Subscriber Identity Module. The card used initially just in GSM phones, but now more widely used to carry the personal subscriber information. Swapping the SIM card from one phone to another enables users to swap phones while retaining their number and other personal information.

SINAD This is similar to the signal-to-noise ratio, but includes other forms of unwanted signal. It is actually the ratio of signal + noise + distortion to noise + distortion. Like the signal-to-noise ratio it is widely used as an indication of signal quality, and under specified conditions it can be used as a measure of receiver performance.

SMS Short Message Service. A service initially used on GSM that enables messages up to 160 characters to be sent by one mobile and received by another.

SNR Signal-to-noise ratio. This is the ratio of the signal to the background noise, and is used as an indication of signal quality. Under specified conditions it can be used as a measure of receiver performance. It is sometimes denoted as S/N.

Soft handover (handoff) This is the ability, used in CDMA, of a mobile to be able to handover when it can hear two base stations at once. In this way it can make the connection to the new base station before relinquishing the connection with the previous one.

Softer handover (handoff) A term applicable to cdmaOne/CDMA2000, this is a soft handover between two sectors of the same base station.

TACS Total Access Communications System. The analogue first-generation system widely used within Europe and a number of other countries.

TDD Time Division Duplex. The concept of using timesharing to enable forward and reverse channels to be accommodated on the same channel or frequency.

TDMA Time Division Multiple Access. The concept whereby multiple users share the same channel by being assigned different timeslots. This scheme is used in a number of digital cell phone systems, including GSM, NA-TDMA and PDC.

TIA Telecommunications Industry Association. An industry association based in North America that organizes standards bodies for the telecommunications industry.

TMSI Temporary Mobile Subscriber Identity. A temporary identifier assigned to a subscriber by a network.

UE User Equipment. The term used in W-CDMA for a mobile handset.

UMTS Universal Mobile Telecommunications System. The name given to the third-generation system that has been developed as a migration path for GSM. It is also known as Wideband CDMA (W-CDMA).

Uplink Another term for the reverse link, i.e. the link from the mobile to the base station.

VLR Visitor Location Register. The database that is used to hold information on mobiles that are 'visitors' to the network, i.e. roaming.

Vocoder A voice encoder. A codec used for encoding and decoding voice signals. Vocoders utilize many of the characteristics of speech to encode signals far more efficiently than if they were other forms of signal.

Walsh codes A family of orthogonal codes used in CDMA technology.

W-CDMA Wideband CDMA. The third-generation mobile phone system that has been developed as the migration path from GSM. It is also known as UMTS.

Index

V

W